Frontispiece; *Vélocipédraisiavaporianna*, 1818 (The First Idea for a Motorcycle)

SCIENCE MUSEUM

Motor Cycles

by

C. F. CAUNTER

A HISTORICAL SURVEY

*As illustrated by the
collection of motor cycles
in the Science Museum*

LONDON
HER MAJESTY'S STATIONERY OFFICE
1970

First published 1956
Second edition 1970

SBN 11 290020 8

Contents

ILLUSTRATIONS	page v
INTRODUCTION	xi
CHRONOLOGY	xii

HISTORICAL SURVEY

Chapter 1.	The Pioneers: 1869–1894	1
Chapter 2.	Motor Tricycles and Quadricycles: 1895–1903	8
Chapter 3.	Early Development: 1898–1908	15
Chapter 4.	Establishment of Basic Principles: 1909–1914	30
Chapter 5.	Interim Development: 1915–1929	46
Chapter 6.	Interim Development: 1930–1945	58
Chapter 7.	Later Development: 1946–1955	66
Chapter 8.	Miniature Motorcycles: from 1956	77
Chapter 9.	Larger Motorcycles: from 1956	91
Chapter 10.	Accessories and Ancillaries: from 1956	98

BIBLIOGRAPHY	109
INDEX	111

Illustrations

Frontispiece: Vélocipédraisiavaporianna: 1818
(The First Idea for a Motorcycle)

PLATE
1 Michaux-Perreaux steam motorcycle: 1869   between pages 2 & 3
   Parkyns-Bateman steam motor tricycle: 1881
   Copeland steam motorcycle: 1885
   Dalifol steam motorcycle: 1894

2 Daimler motorcycle: 1885
   Butler 'Petrolcycle' tricycle: 1887
   Hildebrand & Wolfmuller motorcycle: 1894
   Holden motorcycle: 1897

3 $1\tfrac{3}{4}$ h.p. De Dion-Bouton motor tricycle: 1898   between pages 26 & 27
   $2\tfrac{1}{4}$ h.p. Ariel motor tricycle: 1898
   3 h.p. Renaux motor tricycle: 1899
   $1\tfrac{3}{4}$ h.p. De Dion-Bouton motor tricycle: 1898 (rear view)

4 $1\tfrac{1}{2}$ h.p. Werner motorcycle: 1899
   $1\tfrac{1}{4}$ h.p. Minerva motorcycle: 1901
   2 h.p. Werner motorcycle: 1902
   $1\tfrac{3}{4}$ h.p. Singer motorcycle: 1904

5 $2\tfrac{3}{4}$ h.p. Humber motorcycle: 1902
   $2\tfrac{3}{4}$ h.p. Robinson & Price motorcycle: 1903
   $2\tfrac{1}{2}$ h.p. Triumph motorcycle: 1904
   $3\tfrac{1}{2}$ h.p. Quadrant motorcycle: 1906

6 3 h.p. F.N. 4-cylinder motorcycle: 1905
   $3\tfrac{1}{2}$ h.p. Triumph motorcycle: 1911
   $3\tfrac{1}{2}$ h.p. A.S.L. motorcycle: 1909
   $3\tfrac{1}{2}$ h.p. Indian motorcycle: 1911

7 $3\tfrac{1}{2}$ h.p. P. & M. motorcycle: 1911   between pages 34 & 35
   $2\tfrac{3}{4}$ h.p. Douglas motorcycle: 1911
   $4\tfrac{1}{4}$ h.p. James motorcycle: 1913
   $3\tfrac{1}{2}$ h.p. Rudge-Multi motorcycle: 1915

Illustrations *(Continued)*

PLATE
8 3¼ h.p. J.A.P. engine: 1903   between pages 34
    7 h.p. Wilkinson engine: 1909   and 35
    3½ h.p. Triumph engine: 1912
    2¾ h.p. Douglas engine: 1913

9 2¼ h.p. Velocette motorcycle: 1913   between pages 50
    2¼ h.p. Levis motorcycle: 1916   & 51
    2¼ h.p. Triumph motorcycle: 1914
    2½ h.p. Pullin-Groom motorcycle: 1920

10 4 h.p. Triumph motorcycle: 1917
     3 h.p. A.B.C. motorcycle: 1919
     4/5 h.p. Zenith-Gradua motorcycle: 1920

11 3¾ h.p. Scott engine: 1919   between pages 58
     2¾ h.p. Barr & Stroud sleeve-valve engine: 1922   & 59
     3½ h.p. Triumph-Ricardo engine: 1921
     490 c.c. Norton C.S.1 camshaft engine: 1928

12 348 c.c. Douglas motorcycle: 1928
     348 c.c. Douglas motorcycle: 1947
     487 c.c. Sunbeam motorcycle: 1950
     192 c.c. Velocette L.E. 200 motorcycle: 1953

13 597 c.c. Ariel 'Square Four' engine: 1934
     650 c.c. Triumph vertical twin engine: 1935
     350 c.c. Ehrlich engine: 1946
     197 c.c. Villiers MK. 6E engine: 1952

14 487 c.c. Sunbeam S. 7 engine: 1948
     998 c.c. Vincent 'Rapide' engine: 1951
     192 c.c. Velocette L.E. 200 engine: 1952
     247 c.c. Adler engine unit and transmission: 1954
     123 c.c. Lambretta motor-scooter engine: 1953
     494 c.c. Gilera 4-cylinder engine unit: 1953

15 50 c.c. Raleigh Wisp Cycle Motor: 1967   between pages 90
     49 c.c. Honda 50 Cycle Motor: 1966   & 91
     49 c.c. Suzuki 50 Cycle Motor: 1966

Illustrations *(Continued)*

PLATE
16 88 c.c. Vespa 90 motor scooter: 1966   between pages 90
    148 c.c. Lambretta SX 150 motor scooter: 1966   and 91
    174 c.c. Heinkel Tourist motor scooter: 1960

17 49 c.c. Kreidler Florette Super motorcycle: 1965
    79 c.c. Suzuki 80 motorcycle: 1966
    97 c.c. Yamaha motorcycle: 1966
    152 c.c. Ducati Monza Junior motorcycle: 1965

18 305 c.c. Honda 305 motorcycle: 1966
    248 c.c. A.J.S. motorcycle: 1958
    249 c.c. Norton Jubilee motorcycle: 1952

19 250 c.c. Yamaha 4-cylinder engined motorcycle: between pages 100
    1966   and 101
    250 c.c. Honda 6-cylinder engined motorcycle: 1966

20 745 c.c. Norton Atlas motorcycle: 1967
    1,200 c.c. Harley-Davidson Electra-Glide motorcycle: 1967

21 646 c.c. A.J.S. motorcycle and Watsonian Monza sidecar: 1967

22 Ariel "Square Four" motorcyle (1959)

Diagrams

FIGURE
1. 493 c.c. Isle of Man Sunbeam motorcycle: 1929   between pages 66
2. 499 c.c. Rudge-Whitworth engine: 1921   and 67
3. 500 c.c. oil-cooled Bradshaw engine: 1922
4. 748 c.c. four-cylinder F.N. engine: 1923
5. 2¼ h.p. Sun two-stroke engine: 1922
6. 500 c.c. Dunelt two-stroke engine: 1922
7. 597 c.c. Ariel 'Square Four' engine: 1935
8. 149 c.c. Triumph 'Terrier' engine and unit gearbox: 1954
9. 349 c.c. Royal Enfield engine and gearbox: 1953
10. 498 c.c. Wooler 'Flat-Four' engine and gearbox: 1953
11. 322 c.c. Anzani two-stroke twin-cylinder engine: 1954
12. 49 c.c. Honda engine
13. 249 c.c. B.S.A. Sunbeam motor scooter engine
14. 50 c.c. Ariel 'Pixie' engine   between pages 82
15. 97 c.c. Yamaha twin-cylinder engine   and 83
16. 75 c.c. B.S.A. 'Beagle' engine
17. 247 c.c. Ariel 'Arrow' engine
18. 246 c.c. Yamaha twin-cylinder engine
19. Ducati desmodromic valve gear
20. 344 c.c. Scott twin-cylinder two-stroke engine
21. 499 c.c. B.S.A. twin-cylinder engine
22. 444 c.c. Honda twin-cylinder engine
23. 250 c.c. Norton twin-cylinder engine
24. 498 c.c. Matchless engine

Acknowledgment

Acknowledgment is gratefully made to the Editor of *Motor Cycle*, London, for permission to reproduce in this work the following illustrations:

> Figs. 1-24
> All photographs on Plates 16, 18, 19, 20 and 21
> Honda 50 motorcycle on Plate 15
> Kreidler Florette Super motorcycle on Plate 17

The following are obtainable from the Science Museum:

		Neg. No.
Frontispiece:	Vélocipédraisiavaporianna: 1818	241/54
PLATE 1	Michaux-Perreaux steam motorcycle: 1869	1040/52
	Parkyns-Bateman steam motor tricycle: 1881	243/54
	Copeland steam motorcycle: 1885	1041/52
	Dalifol steam motorcycle: 1894	609/51
PLATE 2	Daimler motorcycle: 1885	454/54
	Butler 'Petrolcycle' tricycle: 1887	501/51
	Hildebrand & Wolfmuller motorcycle: 1894	4542
	Holden motorcycle: 1897	245
PLATE 3	$1\frac{3}{4}$ h.p. De Dion-Bouton motor tricycle: 1898	1363
	$2\frac{1}{4}$ h.p. Ariel motor tricycle: 1898	201/54
	3 h.p. Renaux motor tricycle: 1899	203/54
	$1\frac{3}{4}$ h.p. De Dion-Bouton motor tricycle: 1898 (rear view)	1364
PLATE 4	$1\frac{1}{2}$ h.p. Werner motorcycle: 1899	486/51
	$1\frac{1}{4}$ h.p. Minerva motorcycle: 1901	1042/52
	2 h.p. Werner motorcycle: 1902	538
	$1\frac{3}{4}$ h.p. Singer motorcycle: 1904	2679
PLATE 5	$2\frac{3}{4}$ h.p. Humber motorcycle: 1902	504/38
	$2\frac{3}{4}$ h.p. Robinson & Price motorcycle: 1903	1593
	$2\frac{1}{2}$ h.p. Triumph motorcycle: 1904	507/38
	$3\frac{1}{2}$ h.p. Quadrant motorcycle: 1906	1520
PLATE 6	3 h.p. F.N. 4-cylinder motorcycle: 1905	760/53
	$3\frac{1}{2}$ h.p. Triumph motorcycle: 1911	412/54
	$3\frac{1}{2}$ h.p. A.S.L. motorcycle: 1909	614/51
	$3\frac{1}{2}$ h.p. Indian motorcycle: 1911	780/52

			Neg. No.
Plate 7	3½ h.p. P & M motorcycle: 1911		3328
	2¾ h.p. Douglas motorcycle: 1911		139/54
	4¼ h.p. James motorcycle: 1913		295/39
	3½ h.p. Rudge-Multi motorcycle: 1915		2381
Plate 8	3½ h.p. J.A.P. engine: 1903		642/53
	7 h.p. Wilkinson engine: 1909		2383
	3½ h.p. Triumph engine: 1912		352
	2¾ h.p. Douglas engine: 1913		537
Plate 9	2¼ h.p. Velocette motorcycle: 1913		778/52
	2¼ h.p. Levis motorcycle: 1916		779/52
	2¼ h.p. Triumph motorcycle: 1914		242/54
	2½ h.p. Pullin-Groom motorcycle: 1920		119/54
Plate 10	4 h.p. Triumph motorcycle: 1917		414/54
	3 h.p. A.B.C. motorcycle: 1919		638/53
	4/5 h.p. Zenith-Gradua motorcycle: 1920		510/38
Plate 11	3¾ h.p. Scott engine: 1919		1234
	2¾ h.p. Barr & Stroud sleeve-valve engine: 1922		641/53
	3½ h.p. Triumph-Ricardo engine: 1921		645/53
	490 c.c. Norton C.S.1 camshaft engine: 1928		644/53
Plate 12	348 c.c. Douglas motorcycle: 1928		512/38
	348 c.c. Douglas motorcycle: 1947		246/54
	487 c.c. Sunbeam motorcycle: 1950		318/52
	192 c.c. Velocette L.E. 200 motorcycle: 1953		319/52
Plate 13	597 c.c. Ariel 'Square Four' engine: 1934		7955
	650 c.c. Triumph vertical twin engine: 1935		646/53
	350 c.c. Ehrlich engine: 1946		650/53
	197 c.c. Villiers MK. 6E engine: 1952		636/53
Plate 14	487 c.c. Sunbeam S.7 engine: 1948		649/53
	998 c.c. Vincent 'Rapide' engine: 1951		651/53
	192 c.c. Velocette L.E. 200 engine: 1952		781/52

Introduction

Seventy years of intensive development, large production and wide use have evolved some dozen main forms of motorcycle which provide a great variety of people of all nationalities with convenient, economical and rapid personal mechanical transport. By the middle of the twentieth century, the motor-cycle had become an inherent part of the social structure.

Although Germany and France were the pioneers of the motorcycle before 1900, the ingenuity and industry of the specialist engineers of Coventry, Birmingham and Wolverhampton in particular had, by 1914, established the British motorcycle as the best that could then be obtained. This technical and commercial lead was maintained in the face of growing foreign challenge for the next thirty years. From the 1950's foreign development, in particular that of Italy and Japan, took the lead with remarkably advanced designs having substantially increased specific power outputs and correspondingly higher performances.

The wide adoption of the motorcycle was not due entirely to the facilities made available by its general usefulness. The practical motorcycle served another human need: that of adventure. It made possible to the individual, to a degree unattained by any other form of personal transport, unlimited freedom and scope of road travel and all that this implies. '... this land of policemen and accurate maps and black coats on Sundays,' wrote a veteran motorcyclist,[1] 'is apt to bore a certain temperament. The purchase of a motorcycle imparted a spice of risk and uncertainty and Bohemianism...' And a generation later, an American enthusiast[2] claimed that 'motorcycle riders are the horseback riders of the gasoline age'.

The text of the Historical Survey which follows is intended to serve as an introduction to this story of the development of the motorcycle, with special reference to the National Motor Cycle Collection at the Science Museum, and a separate Catalogue of Exhibits gives detailed technical information about individual items. A list of some of the more important books of reference on this subject, most of which are available in the Science Museum Library, will be found at the end of this book.

Separate chapters are devoted to the main periods of development, and each chapter deals with the motorcycles, accessories and social developments appertaining to its particular period. The National Collection of Motor Cycles illustrates a considerable proportion of the fundamental types, from which have evolved the highly practical and efficient machines of today.

Acknowledgments are gratefully made to the respective proprietors of *The Motor Cycle* and *Motor Cycling* for permission to reproduce certain illustrations.

[1] 'Ixion'. *Motor Cycle Reminiscences*. Messrs. Iliffe and Sons Ltd. 1920.
[2] John Kreitzer. *Saturday Evening Post*. October 30th, 1954, p. 6.

Note. The numbers in the text, e.g. (Inv. 1913–137), are the official inventory identifications of the various exhibits mentioned; these numbers each comprise the year of receipt by the Science Museum and the individual serial number of the exhibit. The official negative numbers of the illustrations included in this book are given on pages ix and x.

Chronology

1818 'Vélocipédraisiavaporianna': The first idea for a motorcycle
1869 Michaux-Perreaux steam motor bicycle
1885 Daimler motor bicycle with medium-speed internal-combustion engine
1888 Pneumatic tyre of J. B. Dunlop
1893 Maybach spray carburettor
1895 De Dion-Bouton motor tricycle with high-speed internal-combustion engine
1895 Simms-Bosch low-tension magneto
1898 Werner front-wheel driven motor bicycle
1901 Minerva, Clement and Singer motor-assisted bicycles
1902 2-h.p. Werner: the first orthodox motor bicycle
1903 Honold-Bosch high-tension magneto
1903 Sidecar patented by W. G. Graham
1904 Werner and Bercley parallel twin-cylinder engines
1904 Fée horizontally-opposed engine of J. F. Barter
1904 F.N. four-cylinder shaft-driven motor bicycle
1904 Truffault swinging-link spring forks
1906 Triumph single-cylinder motor bicycle
1906 Phelon & Moore 2-speed gear and all-chain drive
1906 J.A.P. side and overhead-valve vee-twin cylinder engines
1906 Druid spring forks
1908 Scott twin-cylinder water-cooled two-stroke motor bicycle, with 2-speed gear, clutch and kickstarter
1910 Indian and Wanderer spring frames
1910 Powell and Hanmer motorcycle electric-lighting dynamo
1911 Indian 2-speed countershaft gear and clutch
1911 General adoption of variable-speed gears
1912 Levis and Velocette lightweight two-stroke motor bicycles
1913 James and Sunbeam 'de-luxe' motor bicycles, with 3-speed countershaft gear, clutch and kickstarter and all-chain drive enclosed in oil-bath cases
1914 Adoption of internal-expanding brakes
1914 Indian electric-lighting and engine-starting systems
1919 B.T.H. magdynamo and Villiers flywheel magneto
1919 A.B.C. motor bicycle with horizontally-opposed engine and gearbox unit set across frame, leaf-spring frame, and internal-expanding brakes
1919 A.B.C. 'Skootamota'
1922 Barr & Stroud single sleeve-valve engine
1924 Velocette overhead-camshaft engine
1929 Schnuerle loop-scavenge system for two-stroke engines
1934 Ariel 'Square Four' engine
1935 Matchless 'Silver Hawk' narrow-angle four-cylinder vee engine
1935 Triumph parallel twin-cylinder engine
1936 Adoption of the autocycle
1938 B.M.W. hydraulic cantilever front forks

1938	Gilera four-cylinder motor bicycle
1948	Sunbeam parallel twin-cylinder motor bicycle, with unit gearbox, spring frame and shaft drive
1950	Velocette motor bicycle with water-cooled horizontally-opposed engine, unit-construction gearbox, spring-frame and shaft-drive
1950	Wide adoption of economical motor-assisted bicycles
1960	Wide adoption of the motor scooter and lightweight motorcycle
1965	Rapid progress with high specific output ultra-high r.p.m. single and multi-cylinder engines

Motorcycles

CHAPTER 1

The Pioneers: 1869–1894

Since the idea of the cycle broadly coincided with the early commercial production of the heat engine, it is natural that the possibilities of an engine-propelled cycle should have been considered in the first years of the Industrial Revolution. Moreover, the not inconsiderable experimenting and commercial use which had been achieved with the steam-driven road carriage during the first half of the nineteenth century provided a background of technical development and practical use of mechanical road transport which fostered and aided the evolution of the motorcycle.

This first essentially experimental phase produced a few primitive designs of widely different characteristics which, while they ran in a halting fashion, and a few of the later designs were actually put into small commercial production, were largely tentative essays on how to start designing and building motorcycles rather than serious commercial products.

STEAM MOTORCYCLES

One of the first suggestions for a motorised cycle was made in a cartoon (Frontispiece) of 1818, entitled 'Vélocipédraisiavaporianna' ('A steam-driven velocipede'), which showed one of the two-wheeled velocipedes of hobby-horses, then in vogue, equipped with the suggestion of a steam-engine and boiler unit, but without any technical details to explain how this mechanisation might be carried out. A caption stated that this apparatus was invented in Germany and that an initial trial has taken place in the Luxembourg Gardens, Paris, on the 5th April, 1818. It is unlikely that this machine was built, but the proposal is interesting because it antedates by some fifty years the first motorised cycle which actually ran.

Other suggestions for motorised cycles and coaches appeared during the next decade in the work of various cartoonists, notably Leech, Rowlandson and Cruikshank. It should be remembered that, during the 1820's, steam-driven carriages were running as public transport services on English roads, and it is likely that one or more versions of steam-driven cycles were built during this period. For instance, the steam-driven tricycles in Aitken's prophetic view of Regents Park, drawn in 1831, show certain mechanical details which suggest that they might have been copied from examples which the artist had actually seen. If such vehicles were built, all records of them have been lost, and it is not until 1869 that the first known motorised cycle appeared.

This was basically a Michaux velocipede, known in this country as a 'boneshaker', which was the first commercially successful pedal bicycle. Within its simple frame a particularly neat and compact Perreaux single-cylinder steam-engine unit was installed which drove the rear wheel by means of pulleys and a flexible belt. It was, for its time, quite a practical vehicle and, although it did not lead directly to any commercial development, it is important historically as the prototype of the great variety of motorcycle designs which have appeared since it was made. It is preserved in France in the Robert Grandseigne Collection. (Plate 1.)

Almost contemporary with this machine was the steam-driven boneshaker of the American, S. H. Roper, which is now in the Smithsonian Institution at Washington, D.C., U.S.A.

The next two examples appeared in England. In 1877 a Mr. Meek of Newcastle-on-Tyne made for himself a steam-driven tricycle which is reported to have worked well. Of more importance was the steam-driven tricycle which Sir Thomas Parkyns designed and A. H. Bateman constructed in 1881. This vehicle (Plate 1) consisted of the installation of a neat and efficient steam-engine unit in a contemporary Cheylesmore pedal-tricycle. The condensing water-tube boiler, fired by twenty-one petroleum burners, supplied steam at 200 lb. sq. in. to a small twin-cylinder, double-acting engine which drove one of the road wheels through shaft and bevel-gearing. This tricycle was successfully demonstrated during the Stanley Cycle Show and, as a result, a considerable number of orders were given for it. This initial commercialisation of the motorcycle in England was, however, prevented by the law then existing which made it virtually illegal for such a vehicle to run on the public roads. As the excessive tolls imposed by the law had prevented British pioneer engineers from establishing the steam-driven carriage commercially in the 1830's, so now in the 1880's did the speed restrictions imposed by the Locomotive Acts of 1861 and 1865 effectually prevent other Englishmen from being among the first commercial producers of motorcycles.

This distinction was perhaps achieved by L. D. Copeland of Philadelphia, U.S.A., who, in 1884, fitted a steam-engine unit to an American Star ordinary bicycle. He also produced a steam-driven tricycle which worked sufficiently well to cause some 200 of the type to be built. Those machines satisfied the interest evoked by what was little more than an undeveloped novelty, and there is no record that this first commercial production continued. (Plate 1.)

The French pioneers, Count Albert de Dion (1856–1946) and Léon Serpollet (1858–1907), must also be mentioned as successful constructors of early steam tricycles in 1887 and 1888 respectively, although these vehicles were in both cases experimental essays for the evolution of a series of full-size motorcars, rather than prototypes for the development of motorcycles. Moreover, in the case of de Dion and his partner, Georges Bouton (1847–1938), their major contribution to motorcycle development was to come later and was to be concerned solely with petrol-engined vehicles.

Certain other prototype steam-driven motorcycles were also produced at this time by obscure individuals, some of whose names have been forgotten although their machines are recorded, such as the anonymous steam tricycle 'built prior to 1900' which as formerly preserved in the Motor Museum (1912–1921). In the 1890's, after the first petrol-engined motorcycles had been built, there were a few attempts to produce the steam-driven motorcycle commercially. One of the more important was the Dalifol, of French origin.

An example of the Dalifol steam motorcycle (Inv. 1940–4) is in the National Collections. It was obtained in 1940 from the Newhaven depot of the Southern Railway, where it had been lying abandoned for a great many years. The legend connected with it was that it had taken part in the first London to

Plate 1

Michaux-Perreaux steam motorcycle: 1869

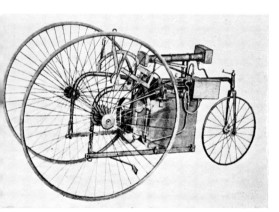

Parkyns-Bateman steam motor tricycle: 1881

Copeland steam motorcycle: 1885

Dalifol steam motorcycle: 1894

Plate 2

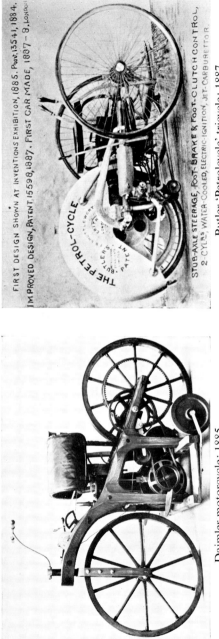

Butler 'Petrolcycle' tricycle: 1887

Holden motorcycle: 1897

Daimler motorcycle: 1885

Hildebrand & Wolfmuller motorcycle: 1894

Brighton run (i.e. the Emancipation Run of 1896) and that it had been left at the depot by a Frenchman. The proximity of Newhaven to Brighton and the fact that No. 18 in the list of starters was a 'French steam bicycle' suggest that this actual machine might well have taken part in that historic event, but there is unfortunately no positive confirmation of this. A similar machine had, the year before, demonstrated its performance by achieving a timed speed of 24·8 m.p.h. (Plate 1.)

The frame of this machine was of approximately safety-bicycle form and carried a steel tubular flash-boiler. Firing was by small-sized coke fuel, stored in the sheet metal container carried within the cycle frame, and was fed by a hopper to the furnace under the boiler. The rear mudguard was in the form of a semi-circular tank which held about five gallons of water. A plunger feed pump, driven by a small overhung crank, supplied water from this tank to the boiler tubes. Control of the power output was obtained by varying the amount of water injected into the tubes by means of a valve directly operated by the rider. The single-cylinder, double-acting, slide-valve engine, supported on the right-hand rear frame stay, operated directly on an overhung crank integral with the wheel hub. The primitive form of telescopic front forks shown was a later modification of the original simple bicycle forks.

This design illustrates the degree of development to which the steam-driven motorcycle had attained in the 1890's, just before the petrol-engined motorcycle appeared in commercial production. When this happened, the steam cycle virtually disappeared and subsequent development was concentrated upon the petrol-engined form of motorcycle.

THE FIRST PETROL-ENGINED MOTORCYCLES

Although a few crude forms of internal combustion-engined carriages had appeared since the beginning of the nineteenth century, there was, apparently, no example of an internal-combustion-engined motorcycle until after Dr. N. A. Otto had achieved the practical application of the four-stroke cycle in 1876. Moreover, the first example of this form of vehicle to be built was equipped, not with the slow speed (250 r.p.m.) type of engine then current, but with an improved type which was lighter in construction and of considerably higher operational speed (800 r.p.m.).

Gottlieb Daimler (1834–1900) was an assistant of Otto in the latter's work on the four-stroke engine, and up to 1882 was managing director of the works at Deutz. After this Daimler devoted himself at his own works at Canstatt to the evolution of a small air-cooled four-stroke horizontal engine (Patent No. 9112 of 1884) which was considerably lighter than previous designs and could operate at speeds up to 800 r.p.m., using either gas or petrol. For the latter fuel, Daimler invented a surface-vaporiser by means of which the engine inhaled air over a broad surface of petrol, thus causing the air and petrol vapour to become mixed in the necessary proportions to form a combustible mixture. Ignition was by means of a hot tube.

In 1885 Daimler improved his original design by the production of a lighter and more compact vertical air-cooled engine (Patent No. 4315 of 1885) which incorporated internal flywheels, fan-cooling, a mechanically-operated exhaust valve, and an automatically-operated inlet valve. This design was the forerunner of the automobile engine as it was subsequently developed. This engine was fitted into a bicycle of Daimler's design (Patent No. 10786 of 1885) which, although crude from the bicycle standpoint, contained some interesting motorcycle features, notably the central upright position of the engine which eventually became standard practice, and the belt-drive to the rear wheel. The bicycle was mostly of wooden construction with light carriage-type iron-tyred wheels, handlebar steering, twist-grip operated brake and a saddle located over the engine. A second version, which was actually built, had the drive taken to a countershaft which connected, by means of a pinion, with an internally-toothed gear ring on the rear wheel. This vehicle was successfully ridden in November, 1886, by Daimler's gifted assistant, Wilhelm Maybach (1846–1929), who was responsible for the invention of the spray carburettor in 1893 and, after 1900, the well-known Mercédès motor cars. The original machine was destroyed in a fire and a replica is (Plate 2) now preserved at the Deutsches Museum in Munich. Daimler gave no further attention to motorcycles, and from then on concentrated upon the development of his motorcar engine.

Almost contemporary with Daimler, an Englishman, Edward Butler (1863–1940), had been working on his own ideas and in 1884 he patented (Specification No. 13541) a design for a motor tricycle, the drawings and description of which (Inv. 1915-404) were exhibited at the Stanley Cycle Show in the same year, and at the Inventions Exhibition in the following year. In 1887 he patented a similar three-wheeled machine, which he called a 'Petrol-Cycle', and which was constructed in the following year. The vehicle had a pair of front steering-wheels and a single rear driving-wheel. Two horizontal, water-cooled cylinders, $2\frac{1}{4}$ in. bore and 8 in. stroke, drove directly on to overhung cranks fixed to the rear-wheel hub. The engine worked on the Clerk two-stroke cycle; a mixture of air and benzoline vapour was exploded in the rear end of the cylinder, the front end forming a pump which compressed the mixture into a reservoir. On the outward stroke of the piston, the mixture was admitted to the cylinder under pressure for about a quarter of the working stroke, and was then fired by an electric spark formed by a wiper breaking contact with the piston. Ignition current was at first provided by Butler's own design for an electrostatic-ignition machine, and later by a Ruhmkorff coil and a battery.

The direct drive proved a failure, as insufficient power was developed at the low engine-speed of about 100 r.p.m. that corresponded with a road speed of 12 m.p.h. Butler, therefore, redesigned the engine in 1889 with a 6 in. stroke and for operation on the four-stroke cycle; moreover, the drive to the rear wheel was transmitted through an epicyclic reduction gear of, at first, 4 to 1 ratio, and later 6 to 1 ratio. This latter provision permitted the engine speed to be increased to 600 r.p.m. As rebuilt, the tricycle in running order weighed 400 lb. (Plate 2.)

In spite of Butler's ingenuity and originality, which had produced various other contributions to technical progress, including a successful design for a spray carburettor which antedated Maybach's design by five years, he was prevented from making a success of his invention by the combined effects of the restrictive road laws at that period and the shortsightedness of his financial backers. Although successful road demonstrations were performed with his tricycle, it was virtually forgotten by 1895 and, in the following year, this first English internal combustion-engined motorcycle was broken up for the scrap value of some 163 lb. of copper and brass that were contained in its structure.

Another interesting English design of this time was produced by J. D. Roots in 1892. This was a front-steering tricycle, which had an original two-stroke single-cylinder oil engine mounted in an inverted position in the rear, which drove the rear axle through bevel and pinion gears. The oil fuel was preheated, and mixed with air; after compression in the crankcase, it was transferred to the cylinder, where it was ignited by a hot tube. Cooling water circulated through the frame tubes of the tricycle. Besides being an early example of the Day crankcase-compression type of two-stroke engine, the Roots tricycle was also important because it foreshadowed the successful Continental motor tricycles which appeared three years later. The Millet bicycle was an interesting and ambitious design with a five-cylinder rotary engine incorporated in the rear wheel. It was manufactured commercially in 1894 by Alexandre Darracq, the pioneer automobile engineer and constructor. An early example (1892) of this design is preserved in the Conservatoire des Arts et Métiers in Paris.

FIRST COMMERCIAL PRODUCTION

What was perhaps the first internal-combustion-engined motorcycle to be produced and sold commercially was designed by the Hildebrand brothers, Henry and Wilhelm, of Munich. In 1889 they made a small steam-engine intending to fit it into a bicycle. In 1892, with the assistance of Alois Wolfmuller and his mechanic, Hans Greisenhof, they made a small two-stroke petrol engine, and in the following year they produced a twin-cylinder four-stroke horizontal petrol engine. This was at first fitted into a safety pedal bicycle, but as this proved to be too weak, a specially designed machine was evolved which they called a 'motorcycle' ('motorrad'), thus originating the name by which this form of vehicle has since become known. This design proved to be sufficiently practical to warrant a comparatively large production in both France and Germany. Some of these machines came to England as a result of a demonstration in Coventry before that city had added the production of motorcycles to its already thriving cycle trade.

The Hildebrand and Wolfmuller machine of 1894 (Inv. 1929–237) had an open frame formed of four horizontal tubes, between which the horizontal engine was mounted, and another four inclined tubes which carried the surface-vaporiser petrol tank and incorporated the steering head. The engine was of the four-stroke type, with two cylinders each of 90 mm. bore and 117 mm. stroke. The pistons were connected by external rods to two parallel cranks on the rear-wheel hub; rubber straps in lieu of a flywheel assisted the idle compression stroke. The cylinders were water-jacketed, and the cooling water was carried in a curved tank which also formed the rear mudguard. A surface vaporiser and hot-tube ignition were used. The machine weighed 115 lb. and was capable of speeds of up to 24 m.p.h. (Plate 2.)

A contemporary English example of the Hildebrand and Wolfmuller type of machine was the motor bicycle designed by Colonel H. Capel Holden, which, like the former, was manufactured for some years in quantity production. It was, in addition, the first four-cylinder motorcycle to be designed and made. This design, which was patented in 1896, was elaborate and original since it incorporated an opposed four-cylinder engine, having each pair of opposing pistons integral, a single gudgeon-pin and two external connecting-rods which acted directly upon overhung cranks fixed to the rear-wheel hub. The first model (Inv. 1910–25) had an air-cooled engine, with a cylinder bore of 2·125 in. and a stroke of 4·5 in. The pistons moved together, but the impulses were arranged to occur alternately in one cylinder at each end so that there was a power impulse at each piston stroke. The surface vaporiser consisted of a tank containing the petrol, with a vertical longitudinal diaphragm of copper wire gauze, through which the air passed on its way to the mixing valve, and part of the exhaust heat was used to warm the fuel to assist vaporisation. Ignition was effected by means of a coil-and-battery system, working in conjunction with a high-tension commutator type distributor, which was driven from the exhaust-valve camshaft. Since there was no rotary motion in the engine itself, a chain-drive had to be arranged from the rear wheel to drive the camshaft. The machine weighed 123 lb. and was capable of a maximum speed of about 24 m.p.h. (Plate 2.)

A later version of the engine (Inv. 1949–180) had water-jacketed cylinder barrels and was capable of producing 3 h.p. at a maximum speed of 840 piston strokes (equal to 420 r.p.m.) a minute. In this form the Holden motor-cycle was produced commercially from 1899 to 1902. It was a well-made machine, and its four-cylinder engine did much to offset the harshness of drive that resulted from the slow speed and the direct coupling of the connecting-rods to the rear wheel cranks.

The young motor industry had, by the middle 1890's, begun to attract the attention of financial adventurers, and one in particular appeared in connection with motorcycles. An American, E. J. Pennington, arrived in this country in 1896 with a motorcycle design of interesting if somewhat dubious features. This machine was essentially a fairly normal diamond frame safety bicycle, with small wheels fitted with large-section tyres. Attached to the rear stays of the frame, and supported by additional diagonal struts, was a simple twin-cylinder horizontal four-stroke engine which drove directly on to cranks attached to the rear wheel hub. The cylinders, of 2½ in. bore and 6 in. stroke, were plain steel barrels with no cooling fins, and a flywheel was built into the rear wheel. The exhaust valves were mechanically operated; the automatic-inlet valves had a common air inlet, over which was a needle valve, operated by the rider, through which petrol dribbled to mix with the incoming air to form a combustible mixture. Ignition was provided by a battery-and-coil system, and a high-tension connection was made to a single-pole plug in each cylinder. This plug carried a spring wiper and each piston crown had a projecting contact in the form of a stirrup. The stirrup engaged the wiper at the top of the piston stroke and spark took place, and on the down-stroke a second spark occurred when the wiper separated from the stirrup. Pennington claimed that this 'long mingling spark' enabled the engine to run on kerosene, since the first spark was supposed to vaporise the fuel and the second spark to ignite it.

Pennington also built a tandem bicycle and a four-seat tricycle on the same lines, and some short demonstration runs were successfully carried out with these machines. None of them went into production, however, although large numbers of orders were received as a result of extravagant publicity. In spite of this, Pennington managed to dispose of his patent rights for about a hundred thousand pounds.

THE LOCOMOTIVES ON HIGHWAYS ACT OF 1896

This pioneer period of motorcycle development terminated in England with the first motor-car trial to be held in this country. In November, 1896, there was organised the historic London-to-Brighton Emancipation Run, so called because it celebrated the repeal of the restrictive 'Red Flag' Road Acts of 1861 and 1865. Until then, the use of engine-propelled road vehicles was virtually impossible in England, and as a consequence development on the Continent, where there was no such restriction, was considerably ahead. The only English-built participant in this event was a $1\frac{1}{4}$ h.p. Beeston tricycle, which was a copy of the De Dion-Bouton tricycle, described in the next chapter, and which was one of the first motorcycles to be made in Coventry.

The Beeston Company also produced a quadricycle on much the same lines, but having two steerable wheels and a seat for a passenger in front; this arrangement of passenger-carrying foreshadowed what later became known as the forecar type of tricycle or quadricycle. A Beeston motor bicycle of 1898, having a $1\frac{3}{4}$ h.p. De Dion-Bouton type of engine which drove the rear wheels through a chain, achieved a speed of 27 m.p.h. on the Coventry cycle track. Another Coventry product of the same year was the Accles tricycle, which was built for the newly-formed British Motor Syndicate founded by H. J. Lawson; this machine was generally similar to the De Dion-Bouton design, particularly with regard to the engine.

These few first designs, although they were based largely upon the progress which was now being achieved, particularly with De Dion-Bouton machines and engines on the Continent, initiated the motorcycle trade in Coventry which was eventually to gain and maintain the lead in the design and production of both cycles and motorcycles. They were, however, crude in design, unreliable in action, uncomfortable and even, particularly in the case of the bicycles which were prone to sideslip, dangerous to ride. This latter disability caused the motor tricycle and quadricycle to be preferred during the next few years.

When the Road Acts, of 1861 and 1865 were repealed and replaced by the Locomotives on Highways Act of 1896, which permitted a speed of 12 m.p.h., the technical development and practical use of engine-propelled road vehicles of all kinds really began in England.

Motorcycles

CHAPTER 2

Motor Tricycles and Quadricycles: 1895–1903

Although the first really practical motorcycles were not to appear until about the turn of the century, there occurred in 1895 a development of fundamental importance upon which subsequent motorcycle engine development directly, and motorcycle frame development indirectly, were based. This was the invention by two Frenchmen, Count Albert de Dion (1856–1946) and Georges Bouton (1847–1938), of the light, high-speed petrol engine as a logical development of Daimler's moderate speed engine of a decade before.

This much improved engine, together with the experience gained during the previous decade in the general design and construction of motorcycles, combined after 1895 to make possible a form of motorcycle which, while it was still primitive in design, heavy and cumbersome in construction, and unhandy in operation, was yet capable of being used successfully in considerable numbers as touring and even racing machines.

THE DE DION-BOUTON ENGINE

Count de Dion had, since 1883, built in conjunction with his partner, Georges Bouton, various steam-driven road vehicles which, as has already been said, established them both among the pioneers of mechanical road transport. In 1895 they turned their attention to the petrol engine, and produced a small single-cylinder, air-cooled four-stroke engine of 50 mm. bore and 70 mm. stroke, which weighed 40 lb. and developed $\frac{1}{2}$ h.p. at the unprecedentedly high speed of 1500 r.p.m. Extensive tests with this engine showed that no undue wear resulted from operation at this speed. The heavy double flywheels were enclosed in an aluminium-alloy crankcase, which was another innovation. Hot tube ignition (soon changed to a coil, battery and contact-maker system) and a surface vaporiser of special design were used. Such was the prototype single-cylinder De Dion-Bouton petrol engine which, with little modification and in powers from $\frac{1}{2}$ h.p. (1895) to 8 h.p. (1902) was to equip various types of De Dion-Bouton motor bicycles, tricycles, quadricycles and motorcars, and be manufactured under licence in various countries, as well as to serve in 1898 as the power units for Louis Renault's first automobile and Alberto Santos-Dumont's first airship. In addition, a large number were also supplied over several years as proprietary units for installation in many different types of vehicles.

THE DE DION-BOUTON TRICYCLE

The prototype De Dion-Bouton engine was installed in a pedal tricycle immediately behind the differential rear axle, which it drove through an open spur gear and pinion. The combined fuel tank and surface vaporiser was placed in the frame under the saddle, and the ignition battery was supported on the top tube. The weight was about 200 lb. The tests of this prototype proved encouraging, and $\frac{3}{4}$ and 1 h.p. versions of the engine were produced in the following year. Three of these tricycles were entered in the Paris–Marseilles–Paris race of 1896, which one of them won at an average speed of more than 14½ m.p.h. Soon after this motor tricycles were produced in ever-increasing numbers, not only at the makers' works at Puteaux, but also under licence in England, Belgium, Germany and America. The De Dion-Bouton tricycle became a fashionable vehicle in Paris, even for ladies, and it was soon to be seen on country roads, demonstrating its capabilities as a long-distance touring machine. The impetus which this practical vehicle and its engine gave to the young motor industry was of the greatest importance.

The De Dion-Bouton tricycle (Inv. 1942-66) of 1898 was fitted with a 1¾ h.p. engine. (Plate 3.) The general design was substantially the same as the 1895 prototype, except for the larger engine and various detail improvements such as the stiff girder-type front forks, separate petrol and oil tanks of large capacity, a stronger frame and an enclosed rear axle. The example made under licence by the Motor Manufacturing Company of Coventry (Inv. 1919-355) was substantially the same, except that it was fitted with a 2¼ h.p. engine. This Company received a contract in 1899 from the War Office for the supply of a number of 2¼ h.p. M.M.C. motor tricycles for use in the South African War; this was the first occasion that motorcycles were used under actual war conditions. In the same year the French army authorities experimented with a number of De Dion-Bouton tricycles, mostly ridden by experienced cyclists or motor drivers during army manœuvres. One of these machines, fitted with a 2¾ h.p. engine, ran in the Paris–Bordeaux–Paris race of 1899 and averaged 28·1 m.p.h. Others were fitted with still larger engines – up to 8 h.p. – and used for track-racing.

The choice of the tricycle as a vehicle for the new engine had been made chiefly because of its inherent stability compared with the earlier motor bicycles, which had been prone to side-slip, particularly on wet and greasy roads, or when attacked by dogs, as then often happened. This consideration of stability, coupled with the practical success of the De Dion-Bouton tricycle, established it as a current type for about five years. In addition to the copies of the De Dion-Bouton tricycle, which were being made in large numbers, new designs on generally the same lines began to appear.

The makes, which were based upon the original De Dion-Bouton design of tricycle in England by about 1899, included the Allard, Enfield and Eadie machines. Substantially the same, but including improved features, were the Ariel (Plate 3) and the Swift. Both these latter machines were exceptionally well-made and were manufactured in considerable numbers; both, moreover, located the single-cylinder engine in front of, instead of behind, the rear-axle, as in the case of the De Dion-Bouton machine. This rearrangement of mass towards the centre of the machine resulted in greater longitudinal stability. In addition, the driving-gears were completely enclosed in a casing as a protection against dust and mud and as a means of ensuring adequate lubrication of the gear teeth. Six of these motor tricycles took part in the notable Thousand Miles Trial of 1900, and all performed with credit. By this time some fifty firms, large and small, were engaged in producing motor-cycles of various kinds in England. The 3 h.p. Renaux design of 1899 (Plate 3) was a French example of this form of motor tricycle.

LATER DESIGNS OF MOTOR TRICYCLES

Attempts were soon made to transform the now well-tried De Dion-Bouton type of motor tricycle into a passenger-carrying machine, and this was successfully achieved in various ways. The first method was to attach an open two-wheeled trailer to the rear. This arrangement served the purpose, although it resulted in a somewhat cumbersome and heavy combination which required considerable pedal-assistance from the rider of the tricycle, since the small engines used had little reserve of power; in addition, the passenger was exposed to the exhaust fumes and the dust raised by the tricycle's wheels. A more practical and satisfactory method was the substitution, quite easily performed, of the front wheel and forks of the tricycle for a front-axle having two wheels, which carried an open seat for a passenger. This arrangement was lighter, more compact and more stable because of the four wheels, and the main disadvantage was the exposed position of the passenger in front. The quadricycle thus formed proved quite popular for passenger-carrying and, moreover, provided a basis for more advanced designs which came to be known as the forecar in both tricar and quadricar forms.

An early example of the tricar was the Humber 'Olympia Tandem' of 1898. This was based upon the pedal tricycle of the same name, and had two steerable wheels in front and a single wheel behind; it was, in fact, a normal safety type of pedal bicycle with the front transformed, in the manner already described, to a three-wheeled form. The passenger sat in an open seat between the front wheels, and the single-cylinder engine was mounted behind the rear wheel on an outrigger and drove the rear wheel by a chain. This design was not successful, but it was revived in 1902 by J. Van Hooydonk, who made the same modification to his Phœnix motor bicycles. This was the basis for the well-known and successful Phœnix 'Trimo' forecars, which were popular until about 1905, when the sidecar replaced them.

Other designs of a heavier and more powerful type were made about this time by the Riley, Singer, Rover, Lagonda, Garrard and Quadrant Companies. These vehicles had engines, some water cooled, of powers up to 6 h.p. for single-cylinder and even 9 h.p. for twin-cylinder units, and overall weights of up to 4 cwt. Some of the more powerful and heavier of these, however, were the forerunners of the later (c. 1910) cyclecars rather than of motorcycles proper. In 1899 the Rover Company produced a motorised bath-chair, which was of normal type, except that it included strong rear-wheels, a power unit which drove the rear axle in the De Dion-Bouton manner, and a seat behind for the driver.

The sidecar, by means of which a normal solo motor bicycle could be easily and quickly converted into a practical passenger-carrying three-wheeler, is supposed to have originated from a cartoon in *Motor Cycling* in January, 1903, although there is reason to believe that the idea was conceived some years previously. W. G. Graham patented the sidecar in January, 1903, and the first models were made by Messrs. Mills & Fulford and placed on the market in the same year. In a few years the sidecar had entirely replaced other forms of motorcycle passenger carrying and, apart from the pillion method, has remained unchallenged for more than half a century.

Although the De Dion-Bouton form of motorcycle was a very considerable advance, particularly in respect of its engine, upon the previous primitive motorcycle designs, yet it had certain deficiencies which eventually caused it to go out of favour in face of competition from new and improved types that were becoming available by about the turn of the century. It was heavy and its limited engine power required considerable pedal assistance on hills and against head-winds; it was comparatively complicated and its open gear transmission was noisy and harsh; its ignition system was unreliable and its surface vaporiser was temperamental.

MOTOR BICYCLES

During this motor tricycle period, just before the end of the century, there were also produced a few motor bicycles which, although still primitive, showed a technical advance upon the first types already described.

One of the first ladies' motor bicycles was made by the Coventry Motor Company in 1897; this was a safety-type ladies' open-frame machine with the rear stays elongated to incorporate an engine which drove the rear tyre by a friction pulley. In the same year the Humber Company also built a ladies' motor bicycle. An elaboration of this idea was the Shaw design, in which the rear tyre was friction-driven through a chain-driven countershaft. Still more advanced was the $1\frac{3}{4}$ h.p. Gibson, which had the engine built into the bottom bracket of the reinforced diamond frame in the manner of the later P. & M. design.

An unorthodox but successful design of this period, and one that formed a precedent for a design half a century later, was the Singer motor-wheel, originally patented by Messrs. Perks & Birch of Coventry in 1899. The aluminium-alloy wheel contained a single-cylinder engine unit and fuel tank and was designed to replace either the front or rear wheel of a pedal-bicycle, or the front wheel of a pedal-tricycle. The engine, positioned on the stationary wheel spindle, drove the wheel-hub through a chain reduction-gear. The last (1904) version (Inv. 1925–19) (Plate 4) had an open-wheel, spoked on one side only, which made servicing simpler. This Singer engine was one of the first motorcycle units to be fitted with a low-tension magneto.

The motorcycle, although still primitive, had now become a reasonably practical vehicle, and as a consequence it began to be commercialised. The Stanley Cycle Show of 1895 was the first to include a motorcycle – a French Gladiator motor tricycle – among its exhibits, and subsequent shows included an increasing number and variety of these new vehicles.

ACCESSORIES

The last decade of the nineteenth century also produced important fundamental developments in the essential accessories required for the functioning of internal-combustion engines in particular, and the operation of mechanically-propelled road vehicles in general. Up to about 1890 these forms of vehicles had depended upon the primitive devices and systems then available such as hot-tube ignition, surface vaporisers and solid rubber tyres, but soon afterwards progress was made towards new and improved systems upon which future technical design and commercial development were to be based.

Ignition

As has already been indicated, some of the first internal-combustion engines developed for mechanical road transport, such as the Daimler, used the simple hot-tube ignition system which was first proposed by A. V. Newton in 1855 (Patent Specification No. 562) for use with primitive gas-engines. Others, including the Benz, Hildebrand and Wolfmuller and Holden engines, employed primitive Ruhmkorff coil-and-battery systems of the general type used by E. Lenoir and others from about 1860, while Butler's first engine of 1887 used an electrostatic friction machine.

The compact coil, battery and contact-maker system, evolved in 1895 by the De Dion-Bouton Company for their successful range of motorcycle and automobile engines, was a considerable advance on previous forms. Soon after its introduction it became standard for a number of years, not only for the many De Dion-Bouton engines that were produced during this period, but also for the majority of others. This system consisted essentially of an induction coil having primary and secondary windings. The primary circuit was energised by a 4-volt accumulator or dry-cell through a contact-maker operated from a half-time engine shaft at the required moment of ignition, thus inducing a high-tension current in the secondary circuit, which was connected to the sparking-plug.

The De Dion-Bouton ignition system, although it was quite practical, had its faults; its contact-maker was difficult to adjust and was liable to breakage and its fragile celluloid-cased accumulator, which was usually housed in a compartment in the sheet-metal tank of the machine, was liable to deterioration and damage through vibration. The inadequacy of ignition systems in general at this time induced F. A. Simms (1863–1949) to review the problem thoroughly and come to the conclusion that what was required was a self-generating electrical machine operated by the engine. As a result, he produced his design for a low-tension magneto (Patent No. 15411/1897), and in 1895 three prototypes (Inv. 1938–568) were made for him by Robert Bosch (1861–1937) of Stuttgart, who since 1886 had been engaged in the manufacture of primitive electrical ignition systems for large gas engines.

The Simms-Bosch low-tension magneto had heavy permanent magnets, a stationary shuttle-wound armature with a single winding, which was located between the magnet pole-pieces, and an oscillating screening sleeve positioned between them. The rapid interruption of the magnetic field by the partial rotation of this sleeve, coinciding with a make-and-break mechanism inside the cylinder, produced the spark for firing the charge. This self-contained unit, in spite of its limitations, was an improvement on previous ignition systems and it began to be used with motorcycle as well as motor-car engines.

The first high-tension magneto was produced by André Boudeville in 1898 and was fitted to an Aster tricycle of the De Dion-Bouton type. Certain deficiencies, such as the lack of a condenser across the contact-breaker points, prevented it from becoming quite satisfactory. It was G. Honold of the Robert Bosch A.G. who perfected the high-tension magneto in 1902 in substantially the form in which it has since been generally used. This system used an armature having superimposed primary and secondary windings;

upon the primary circuit being interrupted by the contact-breaker, the resulting collapse of magnetic flux induced in the secondary winding a hightension voltage which was discharged across the sparking-plug points. The low-tension magneto was used until about 1903, when it and the earlier coil-and-battery system both began to be replaced by the hightension magneto. By about 1907 the latter system was in general use.

Surface Vaporisers and Carburettors

The surface vaporiser, introduced by Daimler in 1884 for his medium-speed engine, presented a substantial surface of petrol over which the induction air was drawn for the purpose of mixing it with petrol vapour in the required proportion of about 15 to 1. Control was effected by means of a throttle governing the air inlet to the vaporiser and another governing the carburetted air inlet to the engine. Instruments of this kind were also used for some of the first motorcycle engines, including the Hildebrand and Wolfmuller, Holden and De Dion-Bouton types. A variation of this method of vaporisation was the wick or gauze type, as used with the later Werner engines, in which the induction air was drawn through wicks or gauzes saturated with petrol. These vaporisers were temperamental in action, since their action was subject to many variables, including the velocity of the air, the volatility and density of the fuel, and the temperature and humidity of the atmosphere. They were consequently difficult to control and needed constant readjustment of the air and throttle valves. They virtually ceased to be used after about 1902.

A much more positive form of vaporiser, and one upon which future development and production were based, was the jet or spray carburettor which was introduced by Edward Butler in 1889. His 'Inspirator', as it was called, drew atomised petrol from a fine jet by means of a depression caused by the velocity of the induction air through a venturi passage. This design, although it worked satisfactorily, was not adopted in practice, and it was the better known design which Wilhelm Maybach introduced in 1893 which became the commercial prototype of this form of carburettor. The design was substantially the same in essentials as has since been generally employed: a float and needle-valve maintained the petrol at constant level in a jet located within a venturi passage, a throttle valve regulated the quantity of carburetted air drawn in by the engine, and an additional air-inlet valve varied the quantity of air passing through the venturi to modify the resulting mixture strength.

Some of the earliest spray carburettors specially designed for motorcycle engines, and based upon the original Maybach design, were of Continental origin, and included the Longuemare, De Dion-Bouton, Krebs and Vaurs types. These instruments, most of which were of the automatic type, were used in England until the introduction of the semi-automatic Brown & Barlow and A.M.A.C. designs in 1903. The first Brown & Barlow carburettor of that year (Inv. 1911–103) was an example of a single-jet instrument; the jet was placed at the bottom of a vertical chamber which was contracted around it, and the main air supply was drawn through holes in the bottom of the chamber. The level of the petrol was kept constant by a float and needle-valve located in an adjacent chamber, to the bottom of which the petrol pipe inlet was attached. An auxiliary air valve was also provided for modifying the mixture strength. Soon after 1903, the spray carburettor was generally adopted for motorcycle engines.

Carburettor and ignition controls were operated by hand levers located on the petrol tank in front of the rider. Improvements in other controls, such as the exhaust-valve lifter to release compression, were introduced soon after the turn of the century.

Tyres

The pneumatic tyre was invented by J. B. Dunlop (1840–1921) in 1888, and was generally adopted by about 1895 for cycles, motorcycles and motor-cars to the virtual exclusion of all other types. Wheels of 22 in. to 28 in. diameter of metal or wood-and-metal construction were used for motor-cycles; tyre sections were small, varying from $1\frac{1}{2}$ in. to about 2 in. Punctures were frequent, due more to the loose flinty roads of the period than to any undue inadequacy in the tyres themselves.

Lamps

Travel by night was quite possible with the motor tricycle and other forms of motorcycle before 1900. The oil and acetylene forms of lamps used were of the more robust types provided for pedal-cycles. It was not until after 1901 that special motorcycle lamps were devised to withstand the excessive vibration and road shocks inherent in motorcycle operation, under which the cycle-type lamps tended to disintegrate. The Lucas firm was an important pioneer in the development of cycle and motorcycle lamps.

Motorcycles   CHAPTER 3
Early Development: 1898-1908

Before the decline and disappearance of the motor tricycle and quadricycle, their drawbacks had begun to turn attention towards other possible lines of development. This reaction was directed, in particular, against their complication and weight and brought about a trend towards mechanical simplicity and lower weight as a means for improving road performance, riding comfort and handling convenience generally.

The appearance of the light and simple Werner motor bicycle, which was designed on radially new lines with a quiet, smooth drive, opened fresh avenues of development and initiated the flood of progressive ideas concerning motorcycle evolution for which the first decade of the twentieth century was notable. During this period, in particular, engines developed in size, power, refinement and flexibility; and frame and accessory design likewise progressed to make the best use of these continually improving facilities provided by the power units. From this considerable experimental and commercial activity, aided by the better techniques of design and manufacture, eventually emerged the basic principles upon which, by 1914, motorcycle design was to become established.

THE WERNER MOTOR BICYCLE

In 1896 two brothers, Michel and Eugéne Werner, of French nationality and Russian extraction, built a motor bicycle with a single-cylinder, horizontal De Dion-Bouton type of engine placed in the rear and driving the rear wheel through a chain and a friction disc. This arrangement was not satisfactory, and the engine was then moved to the steering head of the bicycle frame and a flexible round leather belt was used to drive on to a belt rim on the front wheel. This arrangement proved simple and effective; in particular, the machine was light and easy to operate, and the drive was smooth and quiet. In 1897 the Werner brothers started to manufacture this 'motorcyclette' and produced a dozen of these machines during this year.

These initial Werner machines, although they at once created considerable interest as the first examples of a new idea, suffered from the unreliable tube ignition system with which they were fitted. Moreover, their high centre of gravity made them prone to sideslip on wet or greasy roads and, in the event of a fall, the tube ignition burner usually set fire to the spilled contents of the petrol tank. The De Dion-Bouton type of electric ignition system was soon afterwards adopted, and in this improved form the Werner machines became popular. An 1899 example of this design (Inv. 1949-136) was fitted with a 217 c.c. single-cylinder air-cooled engine of 62 mm. bore and 72 mm. stroke. The drive was taken to a belt rim on the front wheel through a belt of twisted strip rawhide, the ends of which were joined by a detachable fastener. (Plate 4.)

The success of this simple belt-driven machine quickly produced variants of the same general idea. The manufacturing rights were acquired by the Motor Manufacturing Company of Coventry, and the Raleigh Cycle Company also produced a design upon similar lines. The Royal Enfield Company of Redditch produced an interesting variant in which the belt driving the rear wheel was crossed in order to ensure a better grip on the small engine pulley. These machines were the first practical belt-driven motorcycles to be used in this country and they ushered in a form of drive unrivalled for smoothness and simplicity which, in spite of its limitations, was in general use up to 1914.

CYCLEMOTOR UNITS

The first Werner machines were specially constructed motorcycles with the components built in, rather than merely clipped on to the bicycle structure, and as such they are the true ancestors of the modern motorcycle. But the trend towards simplicity at the beginning of the century also made provision of certain proprietary engine units which were specially designed for easy attachment to a standard bicycle frame for the purpose of converting it simply and quickly into a light motor bicycle. These designs were the ancestors of the cyclemotor units which have been so widely adopted since 1945.

The two most popular types were the Minerva and the Clément designs, of Belgian and French origins respectively, and both were made to fit on to a normal bicycle without modification. The 1901 211 c.c. 1¼ h.p. Minerva engine (Inv. 1952–350) was attached by a clamp to the front down-tube of the bicycle frame, and the drive was taken to the rear wheel through a round belt. (Plate 4.)

The 1⅛ h.p. Clément engine unit (Inv. 1952–160) was substantially the same as the Minerva in general principle and differed only in details. The 143 c.c. engine ran at 2000 r.p.m., and was mounted by means of a trunnion bracket which permitted a variety of positions and attitudes for mounting in the frame. The Clément engine of this size was adopted in England just after 1900 by the cycle manufacturer C. R. Garrard. Complete cyclemotor attachment units weighed up to 40 lb., and a speed of more than 20 m.p.h. was possible with bicycles fitted with them.

The majority of these units were of Continental design and manufacture and some went into comparatively large production; for instance, more than 3000 of the Minerva units were made during 1901. An English design of this type was made in 1900 by the British Motor Traction Company (Inv. 1933–404). It followed the general type of the Minerva, except that the engine was smaller and was clipped to the inside of the diamond frame. In other forms, the position of the engine varied considerably, but the location which eventually became general was on the front down-tube, immediately in front of the pedalling-bracket or just inside the diamond frame. A combined surface vaporiser and petrol tank was contained in a flat sheet-metal case, and a second case held the coil and battery; the control levers were all carried on the tank case. The low centre of gravity of these machines provided an advantage over the Werner design, since they were not so liable to sideslip. Pedalling gear was always included and was necessary because of the limited power of the engines.

A compact and logical design for a cyclemotor attachment unit was the Motosacoche, manufactured by H. and A. Dufaux of Geneva. This design

(Inv. 1915–183) was patented in 1901 and continued to be used for attachment to bicycles in various improved versions until 1908, when it was developed as a complete lightweight motorcycle. The 225 c.c. 1¼ h.p. engine, together with the petrol and oil tanks, carburettor and magneto, were contained as a unit in a diamond-shaped sub-frame; this complete unit fitted into the bicycle frame and was attached to it by clamps and thumbscrews. The drive to the rear wheel was by means of a round leather belt. Another cycle motor attachment unit of considerable interest was the Fée, a 2½ h.p. horizontally-opposed twin-cylinder unit designed in 1905 by J. F. Barter.

THE NEW WERNER MOTOR BICYCLE

In the meantime further progress had been made on the Continent which had considerable influence on future development. In 1901 a basically new type of Werner motor bicycle (Inv. 1912–7) appeared which had its 2 h.p. 262 c.c. single-cylinder engine incorporated vertically in the frame structure by means of special lugs which replaced the normal pedal spindle bracket. The aluminium crankcase thus became an integral part of the frame instead of being, as formerly, merely attached to the frame, and this rational blending of engine and frame constituted a considerable advance upon the earlier method. In addition, this new Werner machine incorporated such progressive features as a spray carburettor, hand-pump lubrication, and a foot-brake of the block type operating upon the rear wheel belt rim. The new formula typified by this design was the basis upon which the great majority of motorcycles have since evolved. (Plate 4.)

This rational conception of combining the engine and frame in an integral unit was quickly adopted by many other designers in England as well as on the Continent. One of the first and most interesting forms produced in this country was made by Messrs. Phelon & Moore in 1901; the engine was inclined forward and built into the frame between the steering-head lug and pedal-bracket. The drive to the rear wheel was taken through primary and secondary chains; the pedal-bracket bearings were used to support a countershaft, which had an important additional function when the P. & M. 2-speed expanding clutch gear was included in it in 1905. This compact arrangement of engine mounting and chain-drive was adopted by the Humber Company for their 1902 machine (Inv. 1925–36), which had a 2¾ h.p. 344 c.c. engine inclined and incorporated in the frame in the P. & M. manner, and the pedal-bracket was used for both the countershaft and the pedal-gear spindle. Humber machines of this type were successfully used for both track and road racing as well as ordinary touring purposes. (Plate 5.)

An English design which more closely followed the Werner principles was the 1903 2¾ h.p. 346 c.c. Robinson & Price machine (Inv. 1921–532). The engine was placed vertically in the frame; the crankcase and the bottom brackets were integral and brazed to the frame tubes. The pedal-bracket was incorporated in the frame immediately behind the crankcase, and the engine drove the rear wheel through a V-section belt. (Plate 5.) The Triumph machine of 1904 (Inv. 1938–539) was another example of this early English development on Werner lines; it was a forerunner of the later Triumph machines which were so successful and did so much to foster the motorcycling movement in general. The 2½ h.p. 292 c.c. engine, fitted in the now orthodox position, was the first example of Triumph design and manufacture (Plate 5). Other English companies which were also producing new and progressive designs to the Werner formula, and which may be said to have laid the secure foundations of the English motorcycle industry, included the Rex, Royal Enfield, Raleigh, Singer, New Hudson, Bradbury, James, Bayliss-Thomas, Quadrant (Plate 5), Alldays & Onions and Matchless firms.

DEVELOPMENT OF FOUR-STROKE ENGINES

By the turn of the century the practical if still primitive motorcycle, as exemplified by the products of the De Dion-Bouton Company and later the Werner Company, was established and designs began to multiply and engine powers increase. In addition several new makes of proprietary engines began to appear after 1902 in powers ranging from 1½ to 3½ h.p.

Of the Continental motorcycle manufacturers who benefited by the new techniques evolved by Daimler and De Dion-Bouton at the beginning of the century the more important were: Clément, Peugeot, Gregoire, Antoine, Griffon, Alcyone, Werner, Buchet and Anzani in France; F. N., Minerva, Kelecom and Sarolea in Belgium; Dürkopp, Fafnir, Progress, Pebok and N.S.U. in Germany; Bock & Hollander, Püch, and Laurin & Klement in Austria. The M.M.C., J.A.P., White and Poppe and Simms designs appeared in England. Some of the contributions of these concerns, such as the development of twin- and four-cylinder engines, were of considerable importance to future progress.

The earliest of these designs were to the general De Dion-Bouton single-cylinder specification, having automatically-operated inlet valves, surface vaporisers, coil-and-battery ignition systems and primitive means for lubrication. By about 1903, however, the inlet valve as well as the exhaust valve began to be mechanically operated, spray carburettors and positive hand-pump lubrication systems were increasingly adopted and, after the introduction of the Bosch high-tension magneto in 1903, coil-and-battery ignition systems were increasingly replaced by these compact and reliable units. The clip-on mounting arrangement was retained initially for some of the largest of these engines. The first Triumph motorcycle, which appeared in 1902, had a 2 h.p. Minerva engine clipped to the frame, but by 1904 the engine was built into the frame, which was now designed specifically as a motorcycle. Certain established makers of pedal bicycles adopted the same course, including Royal Enfield, Raleigh, Matchless and others; they used 2¾ h.p. De Dion-Bouton, 2 h.p. Minerva, and 2¾ and 3½ h.p. M.M.C. engines which were now being fitted with Longuemare or De Dion-Bouton spray carburettors in place of the obsolescent surface or wick-vaporisers.

The M.M.C. engines, although built by the Motor Manufacturing Company of Coventry, were of Continental design, and it was not until J. A. Prestwich produced his first motorcycle engine in 1903 that the English proprietary engine came into being. The first design, a single-cylinder air-cooled engine of 293 c.c. capacity, was followed by the first overhead-valve engine (Inv. 1925–192) to be produced in England, which had a bore of

85 mm. and a stroke of 77 mm. (437 c.c. capacity), and both inlet and exhaust valves were positively operated by a single push-pull cam mechanism. (Plate 8.) Other forms and powers of J.A.P. engines followed one another in quick succession, including single-cylinder and vee twin-cylinder side-valve units that rapidly went into large production, as well as several experimental units for use in motorcycles, forecars, cyclecars and even aeroplanes.

The 3 h.p. White and Poppe engine of 1903, a progressive and robust unit, had mechanically-operated valves and, what was then an unusual feature, a well-finned cylinder head. This design was developed for some years until, in 1910, it was adopted by the Ariel Company as their standard engine unit. The 2¾ h.p. Simms engine of 1903 was fitted with mechanically-operated valves, a spray carburettor and a low-tension magneto; it was not so progressive in overall design as the J.A.P. or the White and Poppe engines and was not used to any extent in production. These few designs of about 1903 are of importance in showing the endeavour that the English manufacturers were now making to become independent of Continental design and supply, and to lay the foundations of the home industry. In addition to these proprietary designs, complete English motorcycles having their own engine types began to appear, such as the Bradbury, Humber, Rex, Rover, Royal Enfield, Quadrant and others.

The impetus given to the motorcycling movement in the first years of the twentieth century by Continental development spread not only to England but also to the U.S.A. Here, in addition to various minor concerns, including the Thomas – reputed to have produced the first American motorcycle – Merkel, Yale, Pope, Marsh, Curtiss, Wagner, Royal, New Era and the more important firms of the Hendee Manufacturing Company of Springfield, Mass., and the Harley-Davidson Company of Milwaukee, Wis., were formed. The smaller companies, soon after the turn of the century, were engaged in the small-scale production of what might be called Americanised versions of the c. 1900 Continental types.

George M. Hendee, manufacturer of the Indian pedal-bicycles, became associated in 1900 with Oscar Hedstrom who, in the following year, produced the first 1¾ h.p. single-cylinder Indian motorcycle, which weighed only 98 lb. This was basically a pedal-bicycle with a 1¾ h.p. single-cylinder air-cooled engine incorporated with the saddle tube of the diamond frame. So successful was the performance of this machine that it was put into production in 1902, when 143 were made. This Indian motorcycle at once became popular and, with detail improvements such as the use of steel cylinders machined from solid forgings, it continued to be made in increasing numbers until 1905. A larger single-cylinder model of 2¼ h.p. was produced in 1906 and mechanically-operated inlet valves were adopted for all models in 1908. The 3½ h.p. single-cylinder model appeared in 1908. The Indian twist-grip system of engine controls was introduced in 1904.

The Harley-Davidson Company produced its first motorcycle in 1903. This was a 2 h.p. single-cylinder machine with belt-drive and loop-frame, and improved versions were built until 1909, when a 6 h.p. vee twin-cylinder design was introduced, which initiated the type upon which this concern has since concentrated.

As the single-cylinder engine became more reliable after 1901, the desirability of greater power and smoother running began to be considered. Increased output was at first provided by larger cylinders, and powers began to rise within a year or two from the modest 1¾ h.p. and 2 h.p. of 1900 and 1901 to 2¾ and even 3½ h.p. An alternative way of increasing the power, and also providing increased flexibility and smoothness of running, was to multiply the number of cylinders and make each of relatively small capacity. The vee-twin form, with cylinders set at angles up to 90°, conveniently fitted into the diamond frame by means of the Werner form of engine mounting, which was the most generally adopted at this time for multi-cylinder engines.

By 1905 various examples of vee-twin engine had appeared, which were rated at about 4 to 5 h.p. and were capable of providing both solo motorcycles and motorcycles attached to sidecars with improved performances and smoother running, In particular, designs were developed by Peugeot, Clément and Griffon in France; Hendee and Harley-Davidson in the U.S.A.; N.S.U. in Germany; Püch in Austria; and Minerva in Belgium. The 578 c.c. 4½ h.p. Minerva (Inv. 1952–161) was an example of the progress made by 1906 in the development of the vee twin-cylinder engine on the Continent. This engine was fitted with mechanically-operated valves, magneto ignition, exhaust-valve lifter and spray carburettor, and so foreshadowed the main features of the vee type of engine as it was to develop in large-scale production.

The pioneer of the vee-twin cylinder engine in England was J. A. Prestwich, who, in the few years following 1905, produced a series of designs of various capacities up to 1000 c.c. which proved immediately successful for motorcycles and tricars. The first examples were fitted with automatic-inlet valves, but mechanically-operated side-by-side valves were soon adopted. In 1906, Messrs. J. A. Prestwich & Company produced an overhead-valve vee engine with two separate push-rods and rockers. The latter engines were at first specially made for racing and speed records; the 1000 c.c. '90-bore' and the large 16 h.p. 2700 c.c. designs were specially remarkable in this category.

The vee-twin cylinder motorcycle engine, as developed by the Indian and Harley-Davidson concerns, was established in America by 1908. The first 3½ h.p. Indian twin of this kind (1905) was formed by adding another cylinder to the single-cylinder model already mentioned; this engine was fitted with automatic-inlet valves and high-tension magneto ignition. In the following year the Indian design was developed to include a torpedo-type petrol tank carried on the top frame tube, instead of the tank originally placed over the rear wheel. In 1909 the twin-cylinder vee-engine of 7 h.p. was placed within the diamond of a looped frame, instead of being built into it, as heretofore, and overhead push-rod-operated inlet valves were used. Thus was established the large vee-twin cylinder form of motorcycle which has remained characteristically American for half a century.

The increasing popularity of track and road-racing produced some special machines by Buchet, Anzani and others, the chief feature of which was the large capacity – in some cases over 2 litres – of their power units. The most successful of this latter type were the 14 h.p. Peugeot-engined machine, which attained a speed of 86 m.p.h. at Brighton in 1905, and the 16 h.p.

J.A.P. vee-twin (120 mm. by 120 mm.) 90° engine, which attained a speed of 90 m.p.h. in 1908. These monsters, however, did not play any significant part in the normal development of the motorcycle. The 3¼ h.p. Werner (Inv. 1954–89) and the 5 h.p. Bercley twin-cylinder designs, which appeared about 1905, had their cylinder bores arranged vertically and parallel to one another. These and some others of the same type had a certain commercial popularity for a short time. Although the type was soon abandoned, these early examples are interesting as the forerunners of a type which has been used on a large scale since 1935.

The opposed twin-cylinder four-stroke engine, with a double-throw 180° crankshaft, has firing periods evenly spaced every revolution, good balance and smooth torque. These characteristics make it particularly attractive as a motorcycle engine unit.

One of the first examples of this form was the Barry rotary engine of 1904, which had a single-throw crankshaft; this was the forerunner of a series of rotary and radial motorcycle engine designs which were produced by C. R. Redrup. In spite of its novel features of separate pumping reservoirs for fresh gas charging, and its low weight of 15 lb., it was not a commercial success.

The horizontally-opposed design produced by J. F. Barter in 1905, originally known as the Fée and later the Fairy, was the forerunner of this form of motorcycle unit. The design was acquired in 1906 by Douglas Motors Ltd. of Bristol and became the basis for the long and successful series of Douglas machines of this type. The 2½ h.p. horizontally-opposed air-cooled Fée engine, with a large outside flywheel, drove the rear wheel through a primary chain, countershaft and secondary belt. These features were also incorporated in the later Douglas machines. A 6 h.p. Fairy motorcycle prototype was produced in 1906, having direct belt-drive. In the following year the first 2¾ h.p. Douglas motorcycle appeared with a direct belt-drive and a frame based on the pedal-bicycle type. In 1908 the frame was designed on motorcycle lines, with spring-forks and horizontal tubes incorporated in the lower part of the frame to accommodate the horizontally-opposed engine in what became characteristic Douglas practice. (Plate 7.)

This technically prolific period of motorcycle development produced a number of four-cylinder engines. One of the first was made in 1903 in prototype form by C. Binks of Nottingham; this engine was a 5 h.p. four-cylinder in-line, air-cooled unit, with a spray carburettor, coil-and-distributor ignition, and an all-enclosed chain-drive to the rear wheel. The 3 h.p. 363 c.c. F.N. four-cylinder in-line engine of 1905 (Plate 6), with high-tension magneto ignition and a spray carburettor, was installed in a machine having shaft- and bevel-gear drive to the rear wheel, an internal-expanding rear brake and telescopic front forks. This machine (Inv. 1952–157) was an immediate commercial success and was made in increasing numbers and improved types during the next twenty years. By 1908 the engine size was increased to 500 c.c. capacity, and several refinements such as a plate clutch and a 2-speed gear were added which made the design more comfortable and flexible.

Other Continental examples of the four-cylinder in-line engine of this period were the Austrian Laurin & Klement and the German Dürkopp. In addition various other multi-cylinder arrangements were tried, including three-cylinder fan and in-line types, and four-cylinder and even eight-cylinder vee types; some of these, however, proved to be of more use as embryo aero engines than as power units for motorcycles.

Water-cooling was employed to a small extent for some special applications such as cooling the cylinder-heads or barrels of heavy-duty engines used with tricars or sidecars. The first Scott two-stroke production engine of 1908 had water-cooled heads. The Green water-cooled cylinder incorporated two small honeycomb radiators, thus making the engine unit self-contained. In general, however, although overheating did occur, aircooling was adequate for motorcycle engines even at this early stage, and the advantages of water-cooling were offset by the additional weight and bulk of the radiator and coolant.

DEVELOPMENT OF TWO-STROKE ENGINES

Although the two-stroke cycle had been used for motorcycle propulsion in the Clerk (pump-compression) form by E. Butler in 1887 and in the Day (crankcase-compression) form by J. D. Roots in 1892, the principle was not applied with any commercial success for this purpose until A. A. Scott (1876–1924) began work on the subject about 1900. He embarked on an experimental period which lasted until 1908, during which he evolved a new and successful form of motorcycle engine and frame which survived in production for nearly half a century. He concentrated upon a compact twin-cylinder vertical three-port two-stroke crankcase-compression engine, having a single central flywheel and single-throw cranks in separate crankcases of small volume, which provided a comparatively high compression-ratio for fresh gas-charging purposes.

His first engine, made in 1901, was of this general form and was used to propel a pedal-bicycle through a simple friction-wheel in direct contact with the front tyre. The next engine of two years later was slightly larger; it was mounted behind the steering-head of a pedal-bicycle frame, and the transmission was by means of a round belt to a countershaft and a friction-pulley in contact with the rear wheel rim.

In 1904 Scott produced two more experimental air-cooled engines on the same lines and the first patents were taken out covering this general design. In 1908 a somewhat larger engine was made which had ball-bearing big-ends, and chain-drive was adopted. This 3½ h.p. engine having air-cooled barrels and water-cooled heads and roller-bearings was installed in an original triangulated duplex open frame; in addition a foot-operated expanding-clutch type of 2-speed gear, all-chain drive, a kick-starter and telescopic front forks were used. This revolutionary design was at first regarded with scepticism, but its smooth, flexible engine, easy means of engine starting, convenient speed-gear and clutch, and stable riding characteristics due to its low centre of gravity, made it a success at hill climbs and trials, and soon assured its acceptance for both competition and touring use.

The essential simplicity of the crankcase-compression type of two-stroke engine of low power made it particularly suitable for use as auxiliary power units for cycles, and a few designs appeared on the Continent before Scott's first production motorcycle appeared. A couple of two-stroke motorcycle engines of different characteristics were produced in 1902 by Continental designers and used in this country for a short while. The 1 h.p. Ixion was a small two-port type of engine, the inlet to the crankcase from the carburettor being controlled by synchronising ports in the hollow crankshaft and the main bearing housing. The unit was attached to the steering-head of a pedal-cycle and drove the front tyre through a friction roller. The 2 h.p. Bichrone was of the Clerk or separate pump type of two-stroke engine, having a single working cylinder and a separate pumping cylinder for supplying the former with fresh gas. Later examples included the Austrian Scheibert unit of 1907, which was mounted on the steering head of a bicycle and drove the front tyre by means of a friction roller, and the Brée, which was mounted in the diamond of the frame and drove a large sprocket on the pedal-spindle by a chain. A 130 c.c. capacity two-port two-stroke engine, having an automatic inlet valve for admission to the crankcase, was produced by R. Treskon of Schonebeck, Germany, and a few were sold in this country about 1908 under the name of the Elgin engine unit.

DEVELOPMENT OF THE SINGLE-CYLINDER MOTORCYCLE AFTER 1905

Notwithstanding the many prototype and production forms of engine which were tried during this prolific design period, the single-cylinder four-stroke engine, with its essential simplicity, robustness and inherent reliability, remained the most satisfactory all-round type. The increase in engine capacity up to 500 c.c., combined with the added facilities of efficient spray carburettors and high-tension magneto ignition, now enabled this class of engine to develop into a unit of all-round versatility. By 1905 the typical $3\frac{1}{2}$ h.p. single-cylinder machine, with direct belt-drive, was capable of long journeys with comparative reliability; it had a maximum speed of about 50 m.p.h., a petrol consumption of more than 100 miles a gallon, and it could be purchased for under £50. The majority of manufacturers after 1905 included at least one machine of this general type in their regular production. It was essentially a fixed gear, clutchless machine which necessitated a 'run-and-vault' start and some adroitness to control it through traffic and up steep hills with a $4\frac{1}{2}$ to 1 gear ratio. It was economical and reliable; moreover, it had a character of its own, which appealed particularly to the younger, more agile rider, and the type was retained in production for some years after 1914, when the more flexible type of machine with gears, clutch and kick-starter had become standardised.

The single-cylinder motorcycle was being produced by about 1907 in various powers from $2\frac{1}{4}$ h.p. (250 c.c. capacity) up to $3\frac{1}{2}$ h.p. (500 c.c. capacity) by such manufacturers as Bradbury, Ariel, Brown, Kerry, Matchless, Minerva, Quadrant, Rex, Roc, Triumph and others. It was possible to add certain types of 2-speed gear, such as the N.S.U. epicyclic pulley gear or the Roc epicyclic rear-hub gear, to machines of this type, which made them more suitable for use with a sidecar or in very hilly country, but the majority of them were used without these additions.

In 1906 the Triumph Cycle Company, under the guidance of M. J. Schulte, introduced a new design which, besides being essentially simple and robust, incorporated such progressive features as a reliable $3\frac{1}{2}$ h.p. engine fitted with ball bearings for the main shaft, high-tension magneto, spray carburettor and a pivoted type of spring fork. The improved version of this machine, which appeared in 1908 with a larger (475 c.c.) capacity engine, at once proved itself to be extremely reliable and efficient, not only for touring but also for racing. This was demonstrated when a Triumph machine of this kind won the 1908 Tourist Trophy Race in the Isle of Man at an average speed of 40·4 m.p.h. over a course of 158 miles. Its forte, however, was reliable touring, and it formed the basis for a rapidly increasing Triumph production for many years. Inv. 1938–540 is an example of the classic form of Triumph motorcycle from this period; this is a 1909 model fitted with a 475 c.c. capacity engine.

Other more original forms of single-cylinder motorcycle were also being developed, such as the all-chain driven P. & M. which, following the layout originated in 1901, incorporated its 3½ h.p. single-cylinder engine as part of the frame and, in 1906, added an efficient expanding-clutch type of 2-speed gear. Messrs. Phelon & Moore were one of the first manufacturers to use a speed-gear as a standard fitting, and their basic design of machine as a whole remained, like the Scott, the same for many years. The Humber (Inv. 1925–36), which was originally based upon the P. & M. layout, with all-chain countershaft drive, later developed, like the Ariel and others, into the more standard belt-driven type.

ENGINE ACCESSORIES

As machines and engines evolved in a great variety of experimental forms during this period, so did accessories and ancillaries develop and multiply as the need for them was created by the ever-increasing scope and reliability of the motorcycle. At the beginning of the century, when the motorcycle was limited in performance and utility, it was practically devoid of those additional items which are required for long journeys, traversing hilly country or for night travel, except for such existing cycle equipment as lamps, brakes, etc., which could be readily adapted. Within a decade, however, the motorcycle was fully equipped with a great variety of specially evolved accessories which contributed substantially to its scope and general utility.

The vital engine ancillaries, consisting of ignition, carburation and lubrication systems, kept pace with engine development during this active period so that they contributed to the growing power output combined with increasing reliability to which engine units were now attaining.

Surface or wick-vaporisers were by about 1903 almost entirely replaced by the smaller, more compact and efficient automatic or semi-automatic spray carburettor. Continental designs were produced by Longuemare, De Dion-Bouton, F.N., Vaurs and Krebs. Although differing considerably in detail, all these depended upon the principle of a depression caused by the intake air passing through a venturi, drawing atomised petrol from a metering jet to form a combustible mixture. This convenient instrument, with its positive action and flexible control, was a considerable advance on the earlier vaporiser forms.

Control was at first effected by means of hand-levers situated on the tank or top frame tube immediately in front of the rider, directly connected by rods to the gas and air throttles. In 1907 these two control-levers were positioned on the right-hand handlebar so that they could be operated by the rider without taking his hands from the handlebars, and the air- and mixture-throttles were actuated through flexible Bowden control cables. This improved method, first introduced on Triumph machines, was quickly adopted and continued until the twist-grip type controls came into general use in the 1920's. Bowden cables were, however, beginning to be used for brakes and exhaust-lifter controls as early as 1903, and the twist-grip throttle control was first introduced on Indian machines in 1904.

English designs of semi-automatic spray carburettors, incorporating separate mixture and air-throttles, began to appear in 1904, notably those produced by Brown & Barlow (Inv. 1911–103) and A.M.A.C., and somewhat later, by the Triumph Company, specially for use with their own engines. The Hedstrom spray carburettor, originated in 1901, was perhaps the first fully automatic instrument produced for motorcycle engines, and was used for many years with Indian motorcycles. The vertical venturi passage passed through the centre of the float chamber, within which was a concentric ring-type float, and the sole control was the throttle valve. The Harley-Davidson motorcycle engine was fitted with the Schebler carburettor, which was used later in considerable quantities for both motorcars and motorcycles.

The De Dion-Bouton type of coil, battery and contact-maker form of ignition, which had served its purpose with comparative satisfaction from the beginning of the De Dion-Bouton engined vehicle, continued to be used until about 1905.

The Bosch high-tension magneto, introduced in 1903, was now a thoroughly reliable and practical instrument. It was compact and completely self-contained; its secondary or high-tension winding was capable of generating a strong spark that was adequate for igniting the charge in the cylinder, and it could be conveniently incorporated with the engine. The Bosch high-tension magneto was one of the more important factors in the development of the internal combustion engine and, by 1907, it had virtually replaced all previous forms of ignition for petrol engines. Other German designs of this type which were also successfully used to some extent were the Eisemann and the U.H. A few British designs existed by about 1910, such as the Ruthardt, later adopted by the C.A.V. Company and put into production, but few of the remainder proceeded past the prototype stage.

The first commercial sparking-plugs, of the De Dion-Bouton type with central electrode and porcelain insulation, were unreliable and needed frequent replacement. Better methods and materials, however, became available with increasing use and experience, and improved and more durable sparking plugs were later introduced by Bosch on the Continent and by the Lodge and Sphinx concerns in this country.

The supply of oil to the engine crankcase, where it was distributed by splash to the working parts, was delivered by means of a hand-operated pump at the rate of about a pumpful every 5 to 10 miles, depending upon speed and load. The oil was eventually consumed by passing up the barrel to the combustion chamber.

In order to minimise the noise of the exhaust, the burnt gases were led into an expansion chamber before release to the atmosphere. Up to about 1909 many machines were fitted with a 'cut-out' to their silencers by means of which the gases could be released before reaching the expansion chamber. This device was intended to reduce back pressure and obtain a slight increase of power for special circumstances, but the additional noise eventually caused cut-outs to be made illegal.

TRANSMISSIONS

Although by about 1908 the performance of the simple single-cylinder single-gear type of machine was satisfactory, even in hilly country, yet it was evident that greater flexibility was required, especially when a sidecar was used. There were also the advantages offered by refinements such as a clutch and means of starting the engine other than by pushing the machine and vaulting into the saddle. Auxiliaries of this kind during this period were, however, few in number, primitive in design and generally inadequate in action, and served merely as first essays in the problems involved.

The simplest means of altering the gear-ratio was provided by the adjustable engine pulley. This was arranged to permit one flange of the pulley to be moved axially on a screw thread in relation to the other so that its effective diameter could be varied. This arrangement was soon standardised on belt-driven machines of the simplest type, and after 1908 the design was elaborated to provide for the gear ratio to be varied by moving one flange of the pulley when in motion.

One of the simplest and most effective of the early speed gears was the 2-speed N.S.U. epicyclic engine pulley (Inv. 1952–161). This device was made in the form of an engine belt pulley, which incorporated a single epicyclic-gear train and was attached in normal manner to the engine shaft. Apart from being reliable, this gear unit was cheap to purchase and easy to fit, and so it was particularly suitable for converting a single-gear belt-driven machine to a geared one.

A few 2-speed epicyclic hub gears also appeared by about 1905, such as the Roc, Vindec and some others. Some examples also included clutches of either the multi-plate or the internal-expanding varieties. Types of countershaft gears included the Fafnir, which had a separate gearbox, and the Anglian, which employed two separate secondary chains running on different ratio sprockets, and means at the countershaft to connect either secondary chain to the primary chain by means of dog clutches. The Chater-Lea. 3-speed countershaft gearbox of 1906 was of the layshaft type and was primarily intended for tricars; it was one of the first of this type of speed gear which ultimately became standard practice for motorcycles.

The P. & M. (1906), Scott (1908) and Enfield (1910) 2-speed gears employed two separate primary driving chains running on pairs of sprockets of different gear-ratio combinations, with separate expanding clutches for connecting either primary chain with the single secondary chain. This general type was simple, reliable and effective and was perhaps the first satisfactory motorcycle gear that appeared; moreover, its essential design also provided the additional faculty of a clutch.

The flexible belt was the form of power transmission from the engine to the driving wheel which equipped the majority of motorcycles for some ten years after the turn of the century. By about 1903 the round twisted-leather

Plate 3

2¼ h.p. Ariel motor tricycle: 1898

1¾ h.p. De Dion-Bouton motor tricycle: 1898 (rear view)

1¾ h.p. De Dion-Bouton motor tricycle: 1898

3 h.p. Renaux motor tricycle: 1899

Plate 4

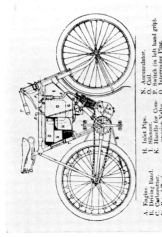

A. Engine.
B. Driving Band.
C. Carburetter.
D. Petrol Tank
E. Vapour Chamber.
F. Air Valve.
G. Gas Valve.
H. Inlet Pipe.
I. Silencer.
K. Handle for Compression Valve.
L. Handle for Contact Breaker.
M. Contact Breaker
N. Accumulator.
O. Coil
P. Switch (in left hand grip).
Q. Interrupter Plug.
R. Sparking Plug.
S. Oil Reservoir.
T. Oil Pump.

THE MINERVA MOTOR BICYCLE.

1¼ h.p. Minerva motorcycle: 1901

1¾ h.p. Singer motorcycle: 1904

1½ h.p. Werner motorcycle: 1899

2 h.p. Werner motorcycle: 1902

2¾ h.p. Humber motorcycle: 1902

2½ h.p. Triumph motorcycle: 1904

2¾ h.p. Robinson & Price motorcycle: 1903

3½ h.p. Quadrant motorcycle: 1906

Plate 6

3 h.p. F.N. 4-cylinder motorcycle: 1905

3½ h.p. Triumph motorcycle: 1911

3½ h.p. A.J.S. motorcycle: 1909

3½ h.p. Indian motorcycle: 1911

belt was replaced by the V-section leather belt which appeared in various forms; the most popular was a built-up variety known as the Watawata. Rubber-and-canvas belts appeared soon afterwards, and eventually became generally used. Apart from simplicity and cheapness, the flexible belt was perhaps the smoothest and quietest form of transmission possible, but it possessed the inherent defects of stretching, excessive slip when wet, particularly on the small-diameter engine pulleys used with single-gear machines, and the pulling out of the fastener. Various other forms of belts were designed to overcome or mitigate these deficiencies, such as leather and chain composite constructions, but without much success.

Chain drive, either direct or through a countershaft, rectified these defects but did not possess the simplicity and smoothness of the belt. Although positive in operation, the harshness of these earlier forms of chain-drive did not encourage their use in this country. They were more used at this period on the Continent and in America.

Shaft and bevel-gear drive, enclosed and self-lubricating, had much to recommend it. Technically, it was fully adequate, and that it was not expensive to manufacture was indicated by the fact that the four-cylinder F.N. motor-cycle with shaft drive was sold for about the same average price as the single-cylinder belt-driven machine.

FRAME ACCESSORIES

As had already happened with the pedal bicycle, the diamond safety type of frame was generally adopted by motorcycle designers after 1902. The engine crankcase was usually attached to lugs on the frame tubes, or less generally carried within a continuous looped frame. The original P. & M. and Indian arrangements of engine mounting involved structural incorporation with the frame; the latter maker abandoned this in 1908 in preference for the loop frame. Certain other designers proceeded on more original lines, such as A. A. Scott, who evolved a completely triangulated form of frame, and F. W. Barnes, who produced the Zenette spring-frame machine.

One of the first examples of spring frame used for motorcycles was the proprietary B.S.A. pedal-cycle design, which was used about 1901 to some extent with Minerva and Clément cyclemotor units. By about 1906 a few more elaborate designs were on the market, including the Sharpe, which employed pneumatic cylinders, and the Bat and Rudge designs which had coil-spring frames.

Spring frames were something of a luxury at this period and were little used, but it was soon realised that spring front forks were necessary as an insulation against road shocks and for better riding comfort and easier control. One of the first and most efficient spring forks was the Truffault design of 1904 which, with its swinging-link action, was the direct forerunner of the type adopted nearly half a century later. The Bat design of 1905 had a simple U-shaped spring-supported carrier for the front wheel.

The Druid fork of 1906 was spring supported and its action was controlled by two pairs of parallel links connecting it with the steering-head; it was a simple yet sound design which became widely adopted. The spring front-forks fitted to the F.N. and Scott motorcycles were among the earliest examples of the telescopic type. The Triumph front fork was pivoted to the steering head of the frame and controlled by a horizontal spring.

Other elaborations of the frame were the addition of a hinged stand, by means of which the machine could be supported when not in use; and a carrier for the transportation of a small amount of luggage.

The stand was generally hinged to the end of the rear fork stays and locked in the down position so that the rear wheel was raised slightly and the machine supported on a firm base; when not in use, it could be swung upwards and clipped to the mudguard. The carrier consisted of a structure of light metal tubes supported from the frame over the rear mudguard, which provided a firm base for the carrying of luggage or, as happened later, to serve as a pillion for a passenger.

The stirrup type of front brake was usually fitted, even after 1912, when machines had become considerably heavier; it was virtually ineffective but was retained chiefly because the law required two independent brakes to be fitted to a motorcycle. Rear brakes, originally of the horseshoe type, developed into the more powerful external-contracting band or the belt-rim block forms, both of which were foot-operated. These were adequate for the speeds and traffic conditions then existing, although they were adversely affected by wet, and they remained the chief forms of motorcycle brake for nearly twenty years. Of particular note is the large and efficient internal-expanding brake incorporated with the rear-wheel hub of the 1905 four-cylinder F.N. machine (Inv. 1952–157). (Plate 6.)

Saddles used for motorcycles were at first of the more elaborate and comfortable kind used for cycles, such as the Brooks B.90. Additional insulation from vibration and road shocks as motorcycles became more powerful and faster necessitated better and more specialised types, and these were supplied by improved designs, intended specially for motorcycle use, such as the Brooks B.100 and B.110. Leather tool-bags of adequate capacity were also adopted, and were generally attached to the carrier.

Higher road speeds and resulting vibration necessitated larger, stronger and more powerful forms of illumination systems. By 1906 the typical motorcycle lamp was of the acetylene type which incorporated a reflector and lens system intended to increase as much as possible the brightness and scope of the illuminating beam.

SOCIAL AND SPORTING ACTIVITIES

Soon after the turn of the century, the future of the motorcycling movement was in some doubt, not only because of technical and financial difficulties inherent in the development of a new industry, but also because of a strong adverse public opinion which, guided by a generally hostile Press, held the motorcycle to be dangerous, dirty and noisy, without considering that it was also, like the motorcar, a social development of considerable potential utility. By 1906, however, the motorcycle was finally established by the advent of really reliable and economical machines, such as the 3½ h.p. Triumph, and the introduction of specialist motorcycling journals which served to inculcate the technical principles, operational procedure and also the ethics of the movement. *Motor Cycling* was started in 1902 by the Temple Press Ltd. and, after a gap of six years, restarted in 1909. *The Motor Cycle* appeared in 1903 under the direction of Messrs. Iliffe & Sons Ltd. These have since become the national specialist publications devoted to the development and use of the motorcycle. The number of licensed motorcyclists increased from 29,000 in 1905 to 50,000 early in 1907. In addition, the Road Act of 1903, while it did not relieve motorists in general from what was considered 'police persecution', assisted the movement by raising the speed limit to 20 m.p.h. and conceding a further degree of acceptance by the authorities of the motoring movement. Registration numbers were also introduced under its provisions.

The formation of certain clubs and associations also served to foster the movement and provide it with a social and even a national status. The first of these, the Auto-Cycle Club, was founded in 1903; in 1907 the name of this organisation was changed to the Auto-Cycle Union, under which it has continued to play a large part as the official governing body which organises the many sporting and social events connected with the movement. The Motor Cycling Club was founded in 1904 as a private members' organisation for the initiation of motorcycling events, such as the London–Edinburgh trial (1904) and the London–Land's End trial (1908). Other major events, such as the Land's End–John o' Groats record, and the English and the Scottish Six-Day Trials were organised by the A.-C.U., as well as many minor events.

Motorcycles had taken part with considerable credit in some of the classic Continental long-distance road races about the beginning of the century, in particular the Paris–Madrid Race of 1903, and as a result a variety of minor events on cycle tracks, test hills and even road circuits became frequent and popular occurrences. Events such as the Land's End–John o' Groats record, which in 1908 stood at 41 hrs. 28 mins., and the maximum speed record, which in the following year was established at nearly 90 m.p.h., indicated the improvement in design and performance which was being achieved. The opening in 1907 of the Brooklands banked race track, upon which motorcars and motorcycles could be driven continuously at well over 100 m.p.h., and the organisation of the first Tourist Trophy Race in the Isle of Man, which was won at an average speed of 38·22 m.p.h., both created invaluable means for the technical development of the motorcycle under the most strenuous operational conditions.

The appearance of motorcycles in the annual Stanley Cycle Show continued until 1905. After this they were included in the Olympia Motor Show.

Motorcycles CHAPTER 4
Establishment of Basic Principles: 1909-1914

The six years immediately preceding 1915 constitute what is perhaps the most important phase in the development of the motorcycle, because during this period were evolved and consolidated the broad essentials of motorcycle design which have formed the basis upon which subsequent progress has been founded. By the end of 1908 the most popular and successful machines were of the simple $3\frac{1}{2}$ h.p. single-cylinder belt-driven variety; by the end of 1914 advanced machines such as the $3\frac{1}{2}$ h.p. Sunbeam with 3-speed counter-shaft gearbox, kick-starter, hand-operated clutch and totally-enclosed all-chain drive were available as standard products.

Apart from the technical development of engines, transmissions, frames and accessories, there were also being evolved certain standard sizes of machine based upon engine cylinder capacity. This method of classification had originated on the Continent for use in racing and hill climbs, with the litre as a unit and fractions of a quarter, one-third and one-half to limit the chosen categories. For the first Tourist Trophy Race, in 1907, the limiting capacities of 500 c.c. for single-cylinder and 750 c.c. for twin-cylinder engines were adopted. By 1911 500 c.c. and 350 c.c. capacities were adopted respectively for the Senior and Junior Races. The 750 c.c. and 1000 c.c. capacities were later standardised for the heavier engines for sidecar use, and 250 c.c. capacity became the upper limit for the lightweight class. The nominal horse-power ratings of these respective capacities – $2\frac{1}{4}$, $2\frac{3}{4}$, $3\frac{1}{2}$, 5/6 and 7/9 – referred to the R.A.C. rating and also, originally, broadly to the actual power outputs developed. But as engines became more efficient and specific outputs increased for the same cylinder capacities, these various categories were eventually identified by their capacities rather than their nominal power ratings. The 475 c.c. Triumph engine of 1908 developed about 4 h.p., while the 499 c.c. Rudge-Whitworth engine which won the 1914 Senior Tourist Trophy Race gave a maximum output of 13 h.p.

Accessories and ancillaries, in particular speed-gears and transmissions, were also developed and elaborated so that they supplemented the maturing design of the motorcycle itself, resulting in more flexible, powerful, comfortable and generally useful machines. This technical progress in turn influenced daily life, as the machines which resulted from it became more and more a part of the social structure. The two or three years immediately before 1915, when human existence was quiet and secure enough to permit the use and enjoyment without undue apprehension as to future world stability of the new facilities which science and engineering were now making possible, were the most propitious period for the normal development and use of this new form of vehicle. Technical progress proceeded with ever-increasing momentum after 1914, but it did so under different and increasingly complex and difficult social and economic conditions.

FOUR-STROKE ENGINE DEVELOPMENT

By 1909 engine design had assimilated the technical achievements of the previous period, and as a result maturer, more efficient units were becoming available in normal production. Engines were produced either as specialist units by a number of established motorcycle firms, such as Triumph, Sunbeam, B.S.A., Premier, Bradbury, Rudge, Norton, P. & M., James and others; or by proprietary firms such as J. A. Prestwich, Peugeot, Minerva, M.A.G., Sarolea, Precision, Burney and Blackburne, Fafnir and others. Some of these proprietary engines were adopted as standard units for a great variety of different makes produced by small manufacturers, of which this period was prolific.

The most popular type of engine was still the simple $3\frac{1}{2}$ h.p. single-cylinder unit, of which the Triumph (Inv. 1912–91) was one of the most representative. (Plate 8.) Retaining the essential features of mechanically-operated valves, internal flywheels, ball and roller main and big-end bearings which its prototype of 1906 had initiated, this engine had an unrivalled reputation for hard work combined with unfailing reliability, without departing from the ideal of extreme simplicity. By 1914 there was a trend to increase the capacity of the single-cylinder engine to make it more suitable for both solo and sidecar use, and the 4 h.p. (557 c.c.) Triumph and B.S.A. models of that year are examples of this class of machine. The 633 c.c. Norton, the 750 c.c. Rudge-Multi, and the 800 c.c. Excelsior 'big-single' models of 1914 were still larger examples of this tendency. Normal crankshaft speeds were in the region of 2500 r.p.m., and compression-ratios were between 4 and 5 to 1; specific power outputs, by reason of improved detail design and manufacturing technique, were increasing, as was demonstrated by the continually improving speeds attained in the Tourist Trophy and other races. The 1911 Senior Race for machines of not exceeding 500 c.c. capacity, which was the first to be held over the arduous Snaefell mountain course, was won by O. C. Godfrey on a $3\frac{1}{2}$ h.p. Indian twin-cylinder machine at an average speed of $47\frac{1}{2}$ m.p.h.; and the 1914 race was won by C. G. Pullin on a $3\frac{1}{2}$ h.p. single-cylinder Rudge-Multi (Fig. 2) machine at an average speed of $49\frac{1}{2}$ m.p.h. In April, 1914, a 490 c.c. single-cylinder side-valve Norton motorcycle attained a maximum speed of 81 m.p.h. and established many world speed records. The Sunbeam designs, which appeared in the 350 c.c. size in 1913 and in the 500 c.c. size in the following year, had such detail improvements as large-diameter valves and heavy-capacity bearings, and an unsurpassed quality of workmanship.

Some single-cylinder designs included innovations to achieve greater efficiency. The Rudge and Indian engines employed a side exhaust valve, with an inlet valve located over it and operated by a push-rod and rocker; this arrangement permitted valves of larger diameter to be used than would have been possible with the normal side-by-side arrangement. J.A.P. and Precision (Inv. 1914–2) engines adopted the overhead arrangement for both inlet and exhaust valves for the same reason and also to achieve a more efficient and symmetrical combustion space. The Premier engine had a small automatically-operated auxiliary exhaust valve located in the cylinder wall and arranged to operate towards the end of the power stroke. Ariel, P. & M., Triumph (Inv. 1926–618), and some other engines incorporated decompressor devices by means of which the exhaust valve was just raised from its seating to relieve the compression pressure and so make engine starting easier. The De Lissa exhaust valve fitted to the Motosacoche engine (Inv. 1915–182) was formed with a double seating which permitted it to be exposed to the air to improve cooling and increase durability. And the 1911 Corah engine employed a combined inlet and exhaust valve of the rotary type.

The single-cylinder four-stroke engine was also developed in lightweight form, with the same progressive specification of the larger units, and with capacities in the vicinity of 200 c.c. Among the more successful of these were the 2½ h.p. Premier, the 2 h.p. Humber, the 2½ h.p. shaft-driven F.N. (Inv. 1938–180), the 2½ h.p. Motosacoche which had been developed as a normal motorcycle from the cyclemotor unit of previous years, and the 2 h.p. Precision. This latter design was of particular interest because of its long vertical valve-operating rockers and its unit construction incorporating a foot-controlled 2-speed gear; it was successfully used during 1914 for a number of lightweight machines which weighed under 100 lb. The 2¾ h.p. Villiers and the 2½ h.p. Veloce engine units were larger examples of this latter formula, and, in addition to a 2-speed gear, also incorporated a clutch.

A large variety of vee-twin cylinder engines of from 250 c.c. to over 1000 c.c. capacities were developed and put on the market during this period. Increased output, together with the smoothness of the multi-cylinder arrangement, made them suitable for a variety of machines from the lightweight solo to the heavy passenger-carrying sidecar machine and even the cyclecar. In the smallest category were the 2½ h.p. Moto Reve, the 2¾ h.p. Royal Enfield, and the 2¾ h.p. Humber; an example of the latter make won the 1911 Junior Tourist Trophy Race at an average speed of 41½ m.p.h. The 3½ h.p. category included such well-designed machines as the Lea-Francis, Zenith, N.U.T., Matchless and A.J.S., which were used mostly for solo riding, although they were capable of pulling a light sidecar. The 5/6 h.p. category was suitable for both solo and sidecar use and included the Bat, Clyno, Royal Enfield, Rex, Sunbeam and Bradbury. The 7/9 h.p. category had, by 1914, become a characteristically American form of motorcycle which was employed for both private use and police-patrol work and included such machines as the Indian, Harley-Davidson, Reading-Standard, Pope, Spacke de Luxe and Excelsior; it was, however, also well represented in England by such designs as the Bat, Matchless, Chater-Lea, Premier, A.J.S. and some others.

A considerable number of manufacturers designed and made their own engines, but proprietary units, such as the J.A.P. and Precision, were also used to a large extent by others; in particular, the robust 680 c.c. side-valve 50° vee-twin unit manufactured by J. A. Prestwich and Co. Ltd. became typical of this form of engine in this country. Continental development in this field was also very active, and was represented by various designs from such established firms as Peugeot, M.A.G., N.S.U., Wanderer, Püch and Minerva.

As a result of the successful commercialisation since 1906 of the Douglas machine employing a 350 c.c. horizontally-opposed twin-cylinder engine, (Plate 7), the easy starting, even torque and good balance characteristics of this form of unit had ensured for it a considerable popularity for light solo machines. The 1912 Junior Tourist Trophy Race was won on a 2¾ h.p. Douglas machine at an average speed of 39½ m.p.h. A similar unit having overhead push-rod operated valves was capable of a maximum speed of over 70 m.p.h.

in 1913. The 3½ h.p. (500 c.c.) Douglas engine appeared in 1914, and advanced designs of the same capacity, having overhead valves, high crankshaft speeds and high specific power outputs, were produced respectively by Granville Bradshaw of A.B.C. Motors and G. Brough. The 500 c.c. A.B.C. established a speed record at Brooklands in 1914 of 80½ m.p.h. The 3½ h.p. Bradbury and Montgomery designs were standard side-valve units of this category. The large 8 h.p. (998 c.c. capacity) Douglas engine of 1912, which was produced both in air and water-cooled forms, was the first of its kind to be made in a large size; it was used with success in the Williamson motorcycle and the Douglas cyclecar of 1913–1914.

Very few designs of the vertical twin type of cylinder arrangement were produced during this period. The experimental 650 c.c. Triumph of 1914 was perhaps the most important, since it was a forerunner of the several designs which have been so widely used since the Triumph Company revived the type for production in 1935. This engine had its crank-throws set at 180° to one another and therefore the disadvantage of uneven firing periods was offset by the advantage of improved mechanical balance. This development of the Triumph Company was stopped partly by the advent of war, and partly because materials and technique were still not adequate to overcome the inherent difficulties which this type presents to the constructor.

A 2¾ h.p. engine of this type was produced in 1909 by the Lloyd Engineering Company, which had the crank throws arranged together, thus permitting even firing periods; the air-cooled cylinder barrels were plain and unprovided with cooling fins.

FOUR-CYLINDER ENGINES

The tendency to use four-cylinder engines in production machines continued to a limited extent and three new designs of advanced features appeared after 1909. Of the earlier types, only the F.N. continued to be produced in improved forms and increased capacities. The model for 1914 was considerably advanced, having a 750 c.c. 'T'-head engine with a separate camshaft arranged on either side of the crankcase to operate respectively the inlet and exhaust valves; a plate clutch was incorporated in the flywheel and the shaft drive contained a 2-speed gear.

The first English machine of this type to be produced commercially was designed by P. G. Tacchi and made by the Wilkinson Sword Company in 1909. The 'L'-head air-cooled engine (Inv. 1922–641) (Plate 8) was of 700 c.c. capacity with mechanically-operated inlet and exhaust valves arranged on one side of the cylinders. The crankshaft was carried in five phosphor-bronze bearings. The power was transmitted from the leather cone-type clutch through a shaft to a 3-speed gearbox, and thence to the spring-mounted rear wheel by means of a universally-jointed shaft and underhung worm gearing. With the low, fully-sprung frame and large bucket type of seat, the smooth torque of its four-cylinder engine, and enclosed shaft and worm-gear drive, the T.A.C.-Wilkinson was one of the most comfortable machines of this period, in spite of its unorthodox appearance.

Two important four-cylinder designs were also put into production in the U.S.A. at this time. The 650 c.c. Pierce-Arrow had an engine of characteristics similar to those of the T.A.C.-Wilkinson, and a multi-plate clutch, 2-speed gear, and shaft and enclosed bevel-gear drive. The engine was mounted in a cradle-type frame constructed of large diameter tubes, which were also arranged to serve as the petrol and oil tanks. This machine was capable of a speed of 60 m.p.h. and, in spite of its size and complication, it weighed only 190 lb.

The 1085 c.c. Henderson four-cylinder engine employed side exhaust-valves and inverted inlet-valves, operated by separate overhead-rockers and push-rods. The three-bearing crankshaft was coupled to the gearbox by bevel gearing, and the final drive to the rear wheel was by chain.

Engines having a radial disposition of cylinders for operation either as rotary or static units have, in spite of the advantage of light weight and smooth torque, played little part in the development of the motorcycle. The limited use that was made of both forms is due largely to the importance they assumed as aero engines during the early years of aviation development. The name of C. R. Redrup deserves mention for the continued effort which he gave to the problem from 1903 to 1920.

His experimental three-cylinder rotary design of 1912, which was made in two versions, having either side valves or concentric poppet valves respectively, was of particular interest. Crankcase and crankshaft were arranged to rotate in opposite directions to obviate unbalanced torque, and both rotating masses drove the rear wheel through separate shaft drives. Ignition contacts were provided through a slip ring.

TWO-STROKE ENGINES

The successful commercial production by A. A. Scott of his classic and original design of two-stroke engined motorcycle did not immediately influence other designers in the same direction. After 1910 Scott produced continually improved water-cooled models which were very successful for hill-climbs and high-speed racing, as well as for solo and sidecar touring. Some of the feats of this unorthodox design at this period were outstanding, such as the 100 consecutive ascents and descents in 1911 of Sutton Bank, which has an average gradient of 1 in 8, in 7 hrs. $31\frac{1}{2}$ mins, and the winning of the Senior Tourist Trophy Races of 1912 and 1913 at an average speed of about 48 m.p.h. These effective demonstrations that the simple three-port crankcase-compression two-stroke engine was capable of sustained use at maximum output did much to establish the Scott design as a sound commercial production, and eventually to influence the adoption of the two-stroke principle by others. (Plate 11.)

The Scott machines up to 1914 followed the prototype design of 1908, as indeed all Scott machines of this type have done for nearly half a century, with the addition of detail improvements. In 1911 both the cylinder barrels and heads were water cooled, in conjunction with an efficient honeycomb type of radiator, and the capacity was increased from 500 c.c. ($3\frac{1}{2}$ h.p.) to 532·5 c.c. ($3\frac{3}{4}$ h.p.). The smooth torque of the engine, the convenience and facility of its controls and its stable riding characteristics, due to the strong rigid frame and low centre of gravity, combined to make it an advanced and practical machine. The special racing machines of 1912–1914 employed delay valves, of either the

2¾ h.p. Douglas motorcycle: 1911

3½ h.p. Rudge-Multi motorcycle: 1915

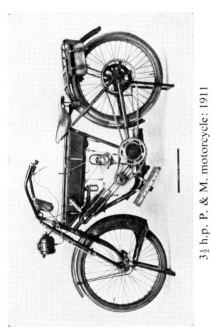

3½ h.p. P. & M. motorcycle: 1911

4¼ h.p. James motorcycle: 1913

D

Plate 8

3½ h.p. J.A.P. engine: 1903

7 h.p. Wilkinson engine: 1909

2¾ h.p. Douglas engine: 1913

3½ h.p. Triumph engine: 1912

rotary or oscillating types, in the transfer passages. These valves permitted the exhaust pressure to fall before the fresh gas was admitted, and also to prolong the opening of the transfer port to assist the efficient charging of the cylinder. Additional scope for the three-port two-stroke engine was eventually found in the lightweight motorcycle, since its essential simplicity and economy as a single-cylinder unit were particularly suited to this form. This lightweight development has already been noted in the case of the four-stroke engine and by 1912, when the Scott had achieved a sound reputation, attention began to be given to the possibilities of the small capacity single-cylinder air-cooled two-stroke engine.

Two firms in particular pioneered this two-stroke lightweight development. Messrs. Butterfields Ltd. produced a simple air-cooled single-cylinder prototype engine of this kind at Stechford, Birmingham, in 1910. This was the basis for the 211 c.c. Levis 'Baby' lightweight motorcycle which was popular during the next decade. It had the simplest specification, with direct belt-drive and an overall weight of under 100 lb., yet its good performance, easy starting, economical consumption (120 m.p.g.) and low price of 33 guineas ensured a large commercial production which did much to establish the lightweight movement. Inv. 1952–176 (Plate 9) shows an example of the Levis 'Baby' two-stroke produced in 1916. Levis engines of $2\frac{1}{2}$ h.p. (293 c.c.) and $2\frac{3}{4}$ h.p. (349 c.c.) were also successfully used in the 1913, 1914, 1920 and 1921 Tourist Trophy Races, and speeds as high as 70 m.p.h. were attained with the largest capacity machines.

The $2\frac{1}{4}$ h.p. 221 c.c. Velocette two-stroke motorcycle (Inv. 1952–91) (Plate 9) which appeared towards the end of 1912 was more elaborate in specification, having an automatic lubrication system, a 2-speed countershaft gearbox, and all-chain drive. An improved version of 249 c.c. capacity was produced during 1914, and used in the Junior Tourist Trophy Race.

By 1913 the two-stroke lightweight machine was established as an economical and reliable alternative to the heavier, more complicated four-stroke machine, and from then new designs appeared in great variety. Several of the large motorcycle manufacturers produced individual designs such as the $2\frac{1}{4}$ h.p. Triumph (Inv. 1947–52) (Plate 9), Royal Enfield, Clyno, Connaught and others; these were mostly fitted with 2-speed gears. The 269 c.c. Clyno engine, in particular, was of unit construction, having the crankshaft geared to the gearbox shaft. In addition, a variety of proprietary engine designs appeared on the market, such as the 269 c.c. Villiers, the 319 c.c. Dalm (Inv. 1953–194), the 350 c.c. Peco and others, which equipped a large variety of new makes of motorcycles selling at prices varying from 25 to 45 guineas. The 350 c.c. Quadrant two-stroke engine employed a poppet-valve controlled exhaust port to effect early opening and closing. The $2\frac{3}{4}$ h.p. (350 c.c.) Wooler of 1912 was another design of considerable originality, which employed a double-acting piston to serve as a combustion head at one end and a pump for fresh gas charging at the other end. The $3\frac{1}{2}$ h.p. Rex and 4 h.p. Roc were examples of large two-stroke engines of English design which appeared as prototypes about 1911. Perhaps the most igenious and interesting of the later two-stroke machines of this period was the 322 c.c. Premier 'Pony' vertical twin-cylinder machine, whose engine employed radial cross-slides in the big-ends to impart a partial rotary, as well as reciprocating motion to the deflector-crowned pistons to achieve more efficient port timings.

This adoption of the small two-stroke engine for motorcycles was not followed to any extent abroad. In 1913, however, the U.S.A. produced the 600 c.c. single-cylinder Schickel, the design features of which included a single-throw overhung crank for the purpose of achieving a crankcase of small volume, a totally-enclosed magneto, and a drive by a flat belt to the rear wheel. This was one of the biggest single-cylinder two-stroke engines that have been used for motorcycles.

CYCLEMOTOR UNITS

The trend towards the development and wide use of lightweight machines also had the effect of reviving the even lighter and simpler method of attaching a small self-contained engine unit to a pedal-bicycle for assistance on hills and against head-winds.

One of the more important examples of this type was the Wall Autowheel, which was, in effect, a self-contained single-wheeled motorcycle designed for easy attachment to a pedal cycle. The first model (1909) incorporated a small horizontally-opposed twin-cylinder two-stroke engine, mounted on a tubular frame which also held the single road wheel. The drive to the 20 in. diameter wheel was through spiral and bevel-reduction gears (6 to 1 ratio). In the following year the engine was redesigned as a 1 h.p. single-cylinder unit, with an outside flywheel; a simple form of 2-speed gear was incorporated in the crankcase of the later models, and the drive from the gear shift was by means of a short chain to the road wheel. In 1912 the Autowheel was redesigned with a 119 c.c. capacity single-cylinder four-stroke engine and, in this form (Inv. 1952–291), it had a considerable use for some years for touring cycles and also as motive power for invalid chairs. In 1915 it was made under licence in the U.S.A. in redesigned form.

A more orthodox type of motor attachment for pedal-cycles was the J.E.S. (Inv. 1927–191), which was manufactured by J. E. Smith of Gloucester. This unit had a 1 h.p. (110 c.c.) four-stroke engine with an automatic inlet valve. The engine was clipped to the front down-tube of the cycle frame, as with the earlier units of this type, and drove through a round leather belt to a rim fixed to the spokes of the rear wheel. The Dayton two-stroke unit was generally similar. A small two-stroke engined unit was produced in about 1913 in the U.S.A. which, mounted on a carrier frame for fixing on the rear of the cycle, drove a small air-screw as a means of propulsion; it was capable of propelling a bicycle at a speed of 30 m.p.h. over good roads.

These auxiliary units added about 40 lb. to the weight of the bicycle, and cost from £12 to £18. They invested the bicycle with a greater range of action and better and easier performance, with a maximum speed of about 25 m.p.h. on level, good quality roads. As such, they must be considered the forerunners of the similar types which, chiefly for economic reasons, became popular utility mounts after 1945.

ENGINE ACCESSORIES

The monopoly which German manufacturers in general, and the firm of Robert Bosch A.G. of Stuttgart in particular, had established soon after

1903 in high-tension magneto-ignition machines continued, so that by 1914 the great majority of internal-combustion engined vehicles all over the world depended upon this source of supply. The technical excellence of the designs as well as the abundant supply had virtually made the development of other designs unnecessary, with the result that this specialised branch of manufacture was neglected in England and other Continental countries. Even in America, a considerable proportion of motorcycles were equipped with Bosch, Eisemann and U.H. magnetos, although American firms such as Splitdorf and Dixie were more independent than Continental firms and developed individual designs.

By 1910 the high-tension magneto, as represented by the Bosch design, had virtually reached its peak of basic development, and was very efficient and reliable in operation, and compact in size. The main developments were the fitting of ball-bearings (1910) and the production of the all-enclosed waterproof type (1912); the latter improvement was of considerable importance since the magneto was usually carried in an exposed position. The Eisemann Company developed the staggered type of armature and pole-piece to equalise the spark intensity provided by magnetos designed to operate vee twin-cylinder engines. The U.H. magneto was fitted with a contact-breaker which operated co-axially with the armature spindle, instead of at right-angles to it as did most of the other designs, and was thus unaffected by centrifugal force.

A few firms tried to foster the art of magneto design and production in this country, notably the Simms, British Thomson-Houston, Thomson-Bennett and the C.A.V. Companies. In spite of their efforts, however, the advent of the war in 1914, resulting in the sudden cessation of the supply of German instruments, found this country in danger of a serious shortage. A source of alternative supply was found in America, and large numbers of Splitdorf and Dixie magnetos were imported. In the meantime, however, strenuous efforts were made to develop the British industry and, by 1916, reliable British ignition instruments were in production.

Although Britain had been dependent in the early years of the century upon foreign designs of carburettors, in particular the Longuemare, by 1909 she had become self-sufficient through the possession of two good semi-automatic designs, the Brown and Barlow and the A.M.A.C., with which, up to 1910, a majority of engines were fitted. Detail improvements were incorporated, such as a taper needle by means of which the jet size was increased in proportion to throttle opening, and an additional rich-mixture jet for easy starting in the B. and B. instrument (Inv. 1914–17); and a multi-orifice diffuser in the A.M.A.C. instrument (Inv. 1921–155).

By 1912 some new types were developed and successfully commercialised. The Senspray carburettor (Inv. 1912–90) had a small subsidiary venturi placed over the jet orifice which augmented and localised the suction caused by the main choke tube, thus effecting a finer atomisation of the fuel. The Cox carburettor, adopted by the B.S.A. Company, had two separate barrel throttles for gas and air control respectively, and a needle-valve by means of which the jet-orifice could be varied by hand. The Binks three-jet instrument (Inv. 1924–723) was designed so that its jets were opened progressively as the ported throttle was lifted from them, thus supplying more fuel as the choke aperture was enlarged. It was claimed to provide a correct mixture strength at all loads and throttle openings to a degree of accuracy beyond the capabilities of the simpler single-jet instrument. The Binks 'Mousetrap' and 'Rattrap' types of carburettors were two-jet instruments, designed with long, variable-aperture, intake passages to provide optimum choke settings for maximum performance. The automatic Stewart Precision carburettor (Inv. 1913–137) was designed to give a constant proportion of petrol to air at all speeds and loads, by means of a suction-lifted air valve controlling the fuel supply; in addition, it also gave a rich mixture for starting. The design of this instrument was particularly compact, having its float chamber formed as an annular space around the choke tube, above which was the air valve.

Refinements in carburettor accessories also began to be adopted, such as needle-valve stop cocks, air-inlet gauzes to prevent the ingress of solid matter, exhaust-heated hot spots in the induction system, and air-lock float-chamber flooding vents by means of which the chamber could be filled to the correct level without wastage.

TRANSMISSION

A considerable number of designs retained the V-section flexible belt for power transmission because of its simplicity and smoothness in action. Its inherent disadvantages of stretching, breaking and slipping in wet weather eventually caused it to be replaced by the chain, although it lasted into the 1920's for single-gear sporting and lightweight machines. In some cases a compromise transition was made in the combined use of primary-chain and secondary-belt drives; this arrangement permitted the belt to operate on a large-diameter driving pulley, and thus lessened the tendency to slip, while at the same time retaining at least some of the flexibility of the belt drive. These belts were now generally made of rubberised canvas, although some leather types such as the Whittle and Service were still used; they were made with a 28 degree section and in widths varying from $\frac{1}{2}$ in. to $1\frac{1}{4}$ in. A great variety of detachable belt-fasteners were also available, but in spite of the simplicity of some and the ingenuity of others, they did little to alleviate the tendency to belt breakage.

Chains for both primary and secondary-drives were, on the other hand, adopted fairly early by some manufacturers, including P. & M., Indian, James, Sunbeam, Scott, Clyno, Royal Enfield and others. In particular, the big vee twin-cylinder type of machine almost invariably adopted this arrangement of drive, in conjunction with a countershaft. Devices known as cushdrives were introduced, particularly by Royal Enfield and Sunbeam, as a means for modifying the essential rigidity of the chain and smoothing out the engine impulses in the drive. The Royal Enfield type had rubber segments in the rear hub through which the secondary chain drove the rear wheel. The Sunbeam type employed a spring-loaded cam-drive incorporated in the engine shaft sprocket which permitted a small relative movement between shaft and sprocket. A further improvement was the adoption of the 'oil-bath' form of chain case by the James and Sunbeam Companies. This arrangement achieved the ideal conditions for chain operation by completely enclosing both primary and secondary chains, and effectively protecting them from wet and dirt. This arrangement also permitted them to operate in a small well of oil, ensuring adequate and continuous lubrication of both chains and sprockets.

The neat and convenient shaft drive was now adopted by a few other designs, the majority of which were of the more elaborate four-cylinder kind. The American Pierce-Arrow four-cylinder machine employed a bevel-gear drive, and the Wilkinson and the American Fielbach used a somewhat more elaborate worm-drive system. In spite of this increase in the use of the shaft for driving purposes, however, the system did not gain any substantial acceptance from manufacturers, and the chain became standard practice by about 1923.

The most important ancillary of the motorcycle which came to be developed and accepted as an essential feature during this period was undoubtedly the variable-speed gear. The increasing scope and performance of motorcycles that were being made available through improved all-round design, together with the wider use to which it was now being put for both touring and competition work, emphasised the limitations of the single-gear machine. In addition, the growing use of the sidecar for passenger-carrying made the need for greater flexibility of control even more apparent.

This situation was generally recognised by about 1910, and, although variable speed-gears existed and were used to some small extent, as has already been mentioned, they were virtually undeveloped and generally unsatisfactory. It was therefore a wise decision that arranged for the 1911 Tourist Trophy Races to be run over the $37\frac{1}{2}$ miles mountain course in the Isle of Man, which abounded in twisting roads and sharp-angled corners, and which varied in altitude from sea-level to over a thousand feet above it. Such a course demanded positive and reliable drives and efficient variable speed-gears, as well as dependable engines and from that time the real development of both these items commenced. The first, second and third places of the 1911 Senior Tourist Trophy Race were taken by Indian machines, fitted with all-chain drives, 2-speed countershaft gearboxes and robust plate clutches. This was the general gearbox and drive arrangement that was ultimately standardised for the majority of subsequent machines.

Before the countershaft gearbox became generally established, however, various other and different types were in considerable use up to 1914, in addition to those earlier devices that have already been described. The adjustable pulley was standard for belt-driven machines, and the compact 2-speed N.S.U. epicyclic-pulley gear continued to be used as a convenient addition. Other designs of pulley gear included the Grado, L.M.C. Auto-Varia and Mabon, which were controlled manually and merely made the engine pulley larger or smaller without provision for belt-tension adjustment. The Philipson pulley incorporated a centrifugal governor, which automatically set the pulley flanges and thus the gear ratio proportionately to speed and load. Some of these devices also incorporated multi-plate clutches.

More elaborate and positive forms of the adjustable-pulley type of gear were the Zenith-'Gradua' (1909) (Inv. 1936–389) and the Rudge 'Multi' (1911) (Inv. 1915–361) designs. The engine-pulley flanges of the 'Gradua' gear were opened or closed by a handle control, and interlocked with this action was a mechanism which moved the rear wheel backwards or forwards, so that, while the effective diameter of the engine pulley was varied, the belt tension was kept constant. The 'Multi' gear opened or closed the engine pulley in step with the closing or opening of the outer belt rim flange on the rear wheel. These gears retained the essential smoothness of the belt and permitted fine changes of gear ratio; they were, however, limited in the range of available gear variation because of the relatively small diameter of the engine pulley on low gear, and they subjected the belts to arduous use so that belt life was usually short. The 'Multi'-gear pulley incorporated a multi-plate clutch, and the later forms of the 'Gradua' gear employed a countershaft which incorporated a cone-type clutch positioned in front of the engine, thus giving the advantage of a large diameter pulley and a long belt drive; both arrangements were, however, only partially satisfactory.

From 1910 to 1914 the form of speed gear most generally used was the epicyclic-hub type, of which various examples were produced. One of the earliest of these was the 2-speed Roc, which either drove through a simple epicyclic-gear train for low ratio or was locked solid for high ratio; the change in ratio was effected by means of pedal-operated external-contracting brake bands. Simple hub-clutches such as the Triumph, which had a considerable number of small-diameter plates, and the Villiers, which had a series of concentric cones to take up the drive and a positive dog-clutch for final engagement, were used to some extent. The B.S.A. 2-speed hub gear also used a clutch of the concentric-cone type. The Harley-Davidson 2-speed hub gear used a bevel differential gear and a system of dog clutches for engagement of the two ratios.

The two most important designs of hub gears at this time, however, were the 3-speed Armstrong-Triplex (Inv. 1911–125) and Sturmey-Archer (Inv. 1920–110) types, the principles of which had been developed for cycle use a decade before. They incorporated multi-plate clutches and were made in various sizes suitable for machines of from $2\frac{1}{2}$ to 8 h.p. They served a useful purpose for a few years but they were in general too weak and complex for motorcycle gears. They were superseded soon after 1914 by the countershaft type of gear.

The 2-speed chain and expanding-clutch type of speed gear, as exemplified by the already established P. & M., Scott and Royal Enfield types, continued to be used during this period. A few 3-speed examples of the latter type, having three primary chains, were also produced for special requirements.

Simple 2-speed countershaft gearboxes, such as were used in the Indian motorcycle in America and the Douglas motorcycle in this country, were already in limited use by 1912. They were of the layshaft type, in which the respective gears were selected and engaged by means of dog-clutches or sliding pinions; plate or cone-clutches could be incorporated in the large driven-sprocket and the addition of a kick-starter mechanism was easy to arrange. The gearbox itself was mounted on the lower rear stays of the frame, with a certain amount of longitudinal movement to permit primary-chain adjustment. More elaborate designs of countershaft gearboxes, which provided 3 and even 4 speeds, became available by about 1913, such as the Clyno,

James, Sunbeam, Juckes and Lea-Francis designs. The 2-speed Bowden gearbox employed expanding clutches, as used on the Royal Enfield and similar gears.

Another important development of gearbox design which was introduced about this time was the unit-construction type, which incorporated the gear unit with the engine crankcase. The primary drive to the gear shaft was usually by means of direct spur-gears or by a noiseless chain. Examples of this form of construction were the $2\frac{1}{2}$ h.p. Veloce, the $2\frac{3}{4}$ h.p. Villiers, the 2 h.p. Precision, the $2\frac{1}{2}$ h.p. Clyno and the $2\frac{3}{4}$ h.p. Diamond designs.

The inclusion of speed gears in shaft-driven machines required a specialised treatment, generally on motorcar lines. The earlier method of incorporating a 2-speed gear in the shaft of the F.N. was replaced in later machines by a motorcar-type 3-speed gearbox mounted immediately behind the flywheel which incorporated a multi-plate clutch. The 2-speed gear and plate-clutch of the Pierce-Arrow design was located in a similar manner.

MOTORCYCLE ACCESSORIES

The rigid diamond type of frame was by now used for the majority of motor-cycles and only detail changes were made to its general form. In particular, by 1912, it was made lower than formerly, with a straight top tube; later this tube was curved downwards at the rear end to permit a still lower saddle position. A number of spring frames also appeared during this period.

Perhaps the best and most successful spring-frame was the Indian which, after 1910, consisted of a strong loop frame, with articulated front and rear forks supported on quarter-elliptic leaf springs. The Wilkinson frame was somewhat on the same principle as the Indian. The Edmund spring-frame consisted of the top tube, which was combined with the saddle-pillar, being pivoted at the steering-head, and supported in rear on a large leaf-spring fixed to the carrier. The German Wanderer and N.S.U. designs employed articulated rear frame forks supported on compression coil springs. The Wooler arrangement mounted both front and rear wheels in small vertically-moving coil-spring units. The Bat design incorporated a combined seat-pillar and footrest attachment, which was hinged to the frame and supported on coil springs, by means of which the rider was insulated from the frame. The most original was the A.S.L. system (Inv. 1948–25) (Plate 6) which incorporated pneumatic buffers for the suspension of both the front and rear portions of the frame; these buffers could be inflated by means of a tyre pump to any required pressure and so provide for any load or road conditions. Both front and rear portions of the A.B.C. frame were suspended on leaf-springs.

In spite of the vibration and road shocks to which motorcycles of this period were subject, the varied and relatively effective spring-frame systems which became available were not used to any extent, except in the case of the Indian machines, whose spring frames and forks were part of the standard specification, and whose production at this time was large.

Spring front forks had become generally adopted by about 1910, and designs were now available in a considerable variety of forms. The simple yet effective girder-type Druid fork, with parallel links and two-coil compression springs, was perhaps the most generally used of the proprietary designs, with the Brampton combined parallel and pivoted action and the parallel-action Saxon types next in popularity. Many manufacturers adopted specialised types of their own, such as the leaf-spring Indian, the pivoted Triumph, and the telescopic Scott, Chater-Lea, F.N., Harley-Davidson and N.S.U. designs. The Bat forks incorporated a pivoted and coil-spring-supported U-shaped arm to which the front wheel spindle was fixed.

In addition to the springing of frames and front-forks, some attention was now also given to the separate springing of saddle-pillars, footrests and even handlebars to provide greater riding comfort. The design of saddles also progressed in the direction of larger seating area of more appropriate anatomical shape, duplex springing and even spring backrests. The Brooks B.150 lightweight and B.170 medium-weight saddles were among the most generally used. Other popular types were made by Lycett and XL-all.

A further contribution to riding comfort, as well as reliability on the road, was the adoption of tyres of somewhat larger section and heavier construction. Tyres of $1\frac{3}{4}$ in. and 2 in. sections were still used for machines having capacities up to 250 c.c. With wheels of 26 in. diameter, a section of $2\frac{1}{4}$ in. was generally used for machines up to 350 c.c. capacity and, sections of $2\frac{3}{8}$ in. or $2\frac{1}{2}$ in. were generally used for machines of larger sizes. In some cases the still heavier 'voiturette' sizes of 650 by 65 mm. and 700 by 80 mm. were adopted for sidecar or heavy touring use. In addition, the treads of these beaded-edge tyres were of heavier construction and had elaborate 'non-skid' patterns for greater safety on wet or greasy roads. The Indian machines used a 28 in. by 3 in. tyre for greater riding comfort, although this policy resulted in a higher riding position. Some tyres had leather treads, or incorporated steel studs and even a multitude of small steel springs to improve durability and riding comfort.

Inner tubes with bicycle-type valves were generally of normal construction, but detachable tubes of the open (Rich) or butt-ended (Dunlop) types, which permitted the easy removal of the tube for repair or replacement without necessitating the removal of the wheel, were also introduced and used to some extent. Sunbeam, Clyno and A.J.S. machines had quickly detachable rear-wheels to permit the quick removal of the inner tube.

The oil lamp had virtually ceased to be used by about 1908, and the acetylene self-generating lamp set began to be particularly used. This type of lamp gave an improved illumination, particularly if used with a Mangin mirror reflector. A considerable improvement was brought about by the introduction by about 1912 of the dissolved-acetylene system, which employed small steel cylinders filled with acetylene gas dissolved in acetone at very high pressure. The gas flow could be regulated positively, and a fresh cylinder fitted with little trouble; the most serious disadvantage of this system was the weight and cost of the cylinders.

The much more convenient alternative of electric-dynamo lighting began to be used in a tentative way by about 1914. Before this, a simple system of electric lighting from dry batteries or accumulators had been used, but the necessity for the continual replacement of batteries or recharging of accumulators limited their use. The first motorcycle lighting dynamo was introduced in 1910 by Messrs. Powell and Hanmer. Small tyre-driven dynamos supplying current directly to the lamp without the use of an accumulator were also used. One or two forms of dynamo, working in conjunction with a separate battery, such as the Mira and the F.R.S. which were generally driven from some convenient point from the engine by means of a belt, were also tried. An arrangement called the 'Maglite' utilised the unused low-tension current of every alternate half-revolution of a magneto fitted to a single-cylinder engine; this current was sufficient to light a single tail-lamp.

The most elaborate motorcycle electric system to be developed and used in production was that fitted as standard to the Indian motorcycles of 1914. This system not only supplied powerful illumination for head, sidecar and tail lights, but also turned the engine over for starting purposes, and supplied current for the ignition coil and distributor. A combined dynamotor was used which was wound either to charge the batteries at 6 volts, or to act as a starting motor at 12 volts.

The increasing scope and reliability of motorcycles as long-distance touring vehicles introduced the need for certain additions in ancillary fittings, in particular speedometers and watches. A considerable variety of accurate speedometers, which indicated road speed, total mileage and individual journey (trip) mileage, were available such as the Jones, Smith and Watford instruments, which operated on a centrifugal governor principle; the Cooper-Stewart instrument, which employed a rotating permanent magnet; and the Cowey, which used an eccentric acting on a circular pendulum. The Isochronous Bonniksen speedometer was more elaborate, having an escapement mechanism which arranged that a pair of hands alternately indicated the speed steadily every five seconds; additional dials recorded miles and tenths of a mile.

In England the speedometer head was generally bracket-mounted on the handlebars, and driven through open gearing and a flexible cable from the front wheel. The mercury inertia-operated Corbin speedometer, fitted as standard to the Indian motorcycles, was mounted on the top frame tube in front of the rider and driven by external gearing and flexible cable from the rear wheel. Watches were sometimes incorporated with the speedometer-head mounting bracket. Motorcycle types of Veeder counters were also available for recording total and trip mileages.

The essential convenience and practicability of the sidecar had, during the ten years following its inception in 1903, established it as the best means of converting the solo motorcyle into a passenger-carrying vehicle. A large variety of types and sizes of sidecars were now available from the light type, having a wickerwork body intended for racing or for light touring with machines of 250 c.c. to 350 c.c. capacity, to the medium and heavy types of coachbuilt forms which, in the larger sizes, were arranged to take two passengers. The sidecar, therefore, fitted to machines ranging from 250 c.c. to 1000 c.c. capacities, was able to convert the motorcycle into a passenger vehicle capable of carrying up to four persons, counting the seat on the carrier. At the outbreak of war in 1914, the sidecar chassis was adapted for various military purposes, notably as a machine-gun carrier.

Brakes were one of the few major items of motorcycle design which, in this country at least, did not make much progress during this otherwise progressive period. The two independent brakes required by law still consisted in general of a pedal-operated block brake acting on the belt rim on the rear wheel, and a cycle-type stirrup brake acting on the rim of the front wheel. These were marginally adequate in dry weather, but when wet they lost much of their limited effectiveness.

Exceptions were to be found in such machines as the P. & M., Clyno, Matchless, F.N. and some others, which fitted brakes of either the internal-expanding or external-contracting band types, and the Indian, which fitted both types on the rear wheel. It was, however, to be another decade before these more positive and reliable types, particularly the internal-expanding variety which could be totally enclosed and so protected from wet and dirt, were generally adopted.

SOCIAL AND SPORTING DEVELOPMENTS

The increasing rate of technical advance and commercial production of a wide variety of motorcycles, combined with a growing popular acceptance of them as sound and reliable vehicles had, by 1911, dispelled the threatened slump in the motorcycle trade and the possible extinction of the motorcycle as a practical vehicle. The Olympia Show for that year contained 275 different models, of which about 80% were fitted with variable-speed gears. The industry expanded under these increasingly favourable commercial conditions and some had, by 1914, become large and flourishing concerns. The Hendee Manufacturing Co. of Springfield, Mass., U.S.A., claimed in 1912 to be the largest makers of motorcycles in the world, having 3000 employees working three shifts a day, and 1800 distributors all over the world. In 1913 there were 55 separate makers in Great Britain, concentrated mostly in Coventry and Birmingham, 36 in the U.S.A., and some 50 on the Continent. In 1910 there were 86,000 registered motorcyclists in this country, and three years later this figure had been doubled. The export trade disposed of 3418 machines in 1910 and, three years later, sold four times this amount at a total value of one million pounds.

The activities of motorcycling clubs also increased. In 1909 the British Motor Cycle Racing Club was formed; the Scottish Six-Days Trial was inaugurated; the John o' Groats–Land's End record time was lowered to 33 hrs. 22 mins.; and the Brooklands track record was raised to 48 miles in one hour. In 1911, the latter record was raised still further to 60 miles in the hour. The world's speed record for motorcycles was established in 1914 at $93\frac{1}{2}$ m.p.h. In 1910 the first London–Exeter run was held in the depth of winter under the worst conditions, and the $37\frac{1}{2}$ miles mountain circuit in the Isle of Man was adopted for the Tourist Trophy Races in the following year.

This initial maturity of the motorcycling movement was also emphasised by a tendency to regard the first commerical machines, of little more than a decade before, as now having historical importance. The Motor Museum, organised by The Temple Press Ltd. in 1912, preserved a few of the very early motorcycles as well as several motorcars; and, in the same year, the Streatham Motor Cycling Club organised the first 'Old Crocks' (*sic*) run from London to Brighton, along the same route taken by the Emancipation Motorcar Run of 1896.

Attention was now being given by the War Office to the potentialities of the motorcycle in war, and some of the motorcyclists of this period became the first Army despatch riders in 1914.

Motorcycles CHAPTER 5
Interim Development: 1915-1929

The advent of a world-wide war in 1914 abruptly arrested the design and production activities which had been so vigorously developing, and the now large motorcycle manufacturing facilities of this country were deflected to various national emergency needs. The Olympia Show, which launched the new products commercially, did not take place in 1914 and the majority of the 1915 designs were made only in prototype form. The technical progress achieved by 1915, therefore, was largely still-born, and only a few of the new designs were put into production for military purposes. The Ministry of Munitions order of 3rd November, 1916, prohibited the manufacture of motor-cycles, except those required for war purposes, and brought to an end the rapid development which had continued since the beginning of the century. Some of these 1915 designs were produced in improved forms when commercial production began again in 1919, but by then different and more difficult conditions had created a new era of motorcycling.

WAR YEARS

Although mechanical transport, including motorcycles, had been used experimentally in the South African (1899–1902) and Balkan (1912) Wars, its use in 1914–1918 was the real beginning of mechanised warfare.

At the outbreak of war, the first military need in connection with motor-cycles was for despatch riders to serve at once with the first seven divisions of the British Expeditionary Force. These men, all enthusiastic volunteers, were trade and trials riders and private motorcyclists from all walks of life. They were mounted upon machines of 1914 type, such as the 4 h.p. belt-driven Triumph, the $4\frac{1}{4}$ h.p. chain-cum-belt B.S.A., the $3\frac{1}{2}$ h.p. all-chain Sunbeam, and the $2\frac{3}{4}$ h.p. chain-cum-belt Douglas. While these served the immediate necessity, a small selection of the 1915 models were put into large-scale production, and used continuously with little modification for the rest of the war. These models included the best known of the despatch rider (D.R.) machines: the Model H Triumph with Sturmey-Archer 3-speed countershaft gearbox (Plate 10), the $3\frac{1}{2}$ h.p. P. & M. used mostly by the Royal Flying Corps, and the $2\frac{3}{4}$ h.p. Douglas (Plate 8) as general-purpose machines. The $3\frac{1}{2}$ h.p. Sunbeam, the $4\frac{1}{2}$ h.p. B.S.A., the $4\frac{1}{4}$ h.p. Norton and a few others were also made in smaller numbers, mostly for use in allied countries. Some types were also used with sidecars for light passenger carrying, such as the $3\frac{1}{2}$ h.p. P. & M. and 7/9 h.p. Indian, or as machine-gun carriers, such as the $3\frac{3}{4}$ h.p. Scott and the 5/6 h.p. Clyno. All these types marked a final transition from the early to an interim stage of development. This progress was shown by such details as the universal adoption of chain-cum-belt or all-chain drives; the countershaft gearbox, incorporating a robust plate clutch and a kick starter; high-tension magneto ignition; semi-automatic lubrication; and, in a few cases, efficient brakes. The Continental manufacturers, who were nearly all directly affected by the European War, were similarly restricted; even the U.S.A. began to modify her production according to European War needs, so that by 1917 improved models, such as the Indian 'Powerplus' and Harley-Davidson designs, were available for use by the U.S. Army.

For the first three years of the war, the production of new designs was largely concentrated in the U.S.A., and some of these were influenced by European and, more particularly, English practice. Soon after this, however, American motorcycles began to decline from the eminent position they had previously held and, after 1919, European and, in particular, English machines had gained a convincing lead. This was largely due to the growing production and use of motorcars by the American nation. During this period of transition, however, the two largest American firms, which had adopted the large vee-twin cylinder machine, developed such models as 500 c.c. horizontally opposed twins and even small single-cylinder lightweight two-strokes, which continued in production for a few years and even served the European market to some small extent. This influence was still further extended when the successful 500 c.c. Indian 'Scout', one of the most successful machines ever produced by this firm, appeared in 1919 to the design of C. B. Franklin, who had for many years taken an active part in the motorcycling movement in England. The Indian 'Powerplus', with a side-valve engine, replaced in 1917 the earlier Hedstrom-designed Indian machines which had made the reputation of the Hendee Manufacturing Company.

Continental design during the war period stagnated and even became retrograde, as was shown by such expedients as the use of spring-spoked wheels as a substitute for rubber pneumatic tyres, which Germany in particular had eventually to adopt under an ever-increasing shortage of suitable materials.

THE NEW ERA

The Olympia Show of 1919 introduced a new era of motorcycle design and production, which was all the more technically prolific and commercially successful because of the enforced production gap of five years. The designs of 1915, the tentative speculations of individual designers during the war, the technical advances achieved for war purposes, particularly in the field of aero design, and the pent-up enthusiasm of would-be users all combined to produce a remarkable variety of new models. In addition, the facilities offered by mass production were becoming generally available so that, although the designs were now more complicated, with overhead-valve engines, dry-sump pressure-lubrication systems, countershaft gearboxes, electric lighting and other advanced features, they were eventually produced at much the same cost as formerly. During the brief period of inflation in 1920, the cost of machines and petrol rose to three and four times what they had been in 1914.

Some two hundred different models, produced by over a hundred firms, were available in 1919, which included such types as the simple single-cylinder belt-driven machine, the two-stroke lightweight, the improved medium-powered single or twin-cylinder machine and the elaborate big-twin sidecar outfit.

DEVELOPMENT OF FOUR-STROKE ENGINES

Although the side-valve single-cylinder machine continued to be popular during this period for normal solo and sidecar use, as it had been during the past decade, advanced aero-engine design influenced the introduction of overhead-valve engines with efficient hemispherical combustion heads, large diameter valves, overlap valve timing, and aluminium-alloy pistons. Specific powers and crankshaft speeds consequently began to rise appreciably, and soon the 4000 r.p.m., high-compression engine began to be produced commercially. Proprietary J.A.P. and Blackburne engines to this specification were the first to be used with success. The larger manufacturers took some time to follow this trend, chiefly because they were already committed to the more standard forms and a sudden change in major production was undesirable. From these prototypes of 1920, however, proceeded the long line of overhead-valve engines which subsequently became standard practice.

It was also the firm of J. A. Prestwich & Co. Ltd. which pioneered the still more advanced overhead-camshaft form of engine as a means for yet higher specific power outputs with increased crankshaft speeds up to 6000 r.p.m. Experimental models of 350 c.c. and 500 c.c. capacities were produced in 1922 for the Tourist Trophy Races and used with some success. Designs of this type were later produced by other firms, such as the 493 c.c. Sunbeam (Fig. 1), the 500 c.c. Norton type C.S.1. of 1928 (Inv. 1928–92) (Plate 11) and the 350 c.c. Velocette of 1924, both of which employed a shaft and bevel gear to drive the single camshaft located on the cylinder head, by which the two valves were actuated. Other designs used a chain-drive to the overhead camshaft.

A more elaborate arrangement in the interests of improved efficiency was the push-rod-operated overhead four-valve design which H. R. Ricardo produced in 1921 for the Triumph Company (Inv. 1921–709) (Plate 11). Each push-rod operated a double rocker-arm which, in turn, operated the pairs of inlet and exhaust valves. The design was first used in the Tourist Trophy Races, and was then produced commercially with success for some years. The efficient combustion head, light valve mechanism and large aggregate port areas enabled this design to produce a high output at high crankshaft speeds. The Rudge-Whitworth engine of this period still employed the inlet over exhaust valves arrangement (Fig. 2).

Two other developments at this time were the Barr & Stroud sleeve-valve engine (Inv. 1923–19) (Plate 11) and the Bradshaw oil-cooled engine (Inv. 1927–1953) (Fig. 3). The Barr & Stroud engine employed a single sleeve-valve of the Burt-McCollom type, which combined a reciprocating with a semi-rotary motion. The engine was simple and worked well within its limits, but as it was made before the principle was established by H. R. Ricardo that piston, sleeve and cylinder-barrel must expand under heat at the same rate, it proved insufficiently reliable. The Bradshaw design (Fig. 3) enclosed the cylinder-barrel within an extension of the crankcase so that a large volume of cool oil could be continually sprayed over it for cooling purposes. Although various forms of these engines were produced for motorcycles and light cars, it is doubtful whether the added complication provided more effective cooling than the normal finned cylinder. Air cooling was now becoming more efficient with increased and better arranged fin area, and the availability of improved materials, such as aluminium-alloy for pistons. The Bradshaw oil-cooled engine perhaps owed its few years of commercial use to its novelty rather than to technical superiority.

In spite of the tendency of other forms of engines to adopt progressive features that promoted improved output and efficiency, the vee-twin form of engine, which was in the main used for sidecar and heavy touring work and therefore tended to remain uncomplicated, continued to be made in the side-valve form, such as the 7 h.p. A.J.S. engine (Inv. 1928–79). The comparatively few examples of this type which incorporated overhead-valves, such as J.A.P., were generally intended for special purposes such as record breaking. A few still more elaborate designs intended for the same purpose, such as the Indian and Harley-Davidson 1000 c.c. track models, employed four valves for each cylinder.

The chief English example of the big twin-cylinder machine at this period was the elaborate Brough 'Superior' of 1925 which, equipped with 1000 c.c. capacity J.A.P. or Anzani engines, was capable of a high performance. A new note in vee-twin cylinder design was struck in 1928 by the 246 c.c. P. & M. 'Panthette' lightweight machine, which had a unit-construction vee-twin engine, the cylinders of which were set across the frame.

The twin-cylinder horizontally-opposed form of engine was now improved in various details. The best known of this type, the Douglas, which had incorporated overhead valves for racing in 1914, now included this valve arrangement in high performance models for commercial production. In addition to improved versions from established firms like Humber and Brough, new designs of engines were produced by Coventry-Victor and Wooler, and, shortly after, the horizontally-opposed form of engine was adopted on the Continent by the German firm of B.M.W. The engine of the B.M.W. was set across the frame and drove the rear wheel through a shaft, a formula which was followed by this concern with increasing success for many years. Mechanically-operated pressure-lubrication systems began to be adopted to provide for the greater stresses and loadings inherent in higher performance.

The most important and progressive machine of this type to appear during this period was undoubtedly the 3 h.p. 398 c.c. A.B.C. (Inv. 1953–325) (Plate 10), designed by Granville Bradshaw in 1918 and manufactured for three years by the Sopwith Aviation Company. The neat and compact engine unit with its overhead-valve cylinders set across the cradle-frame was built on car principles, with an enclosed flywheel and plate clutch and a 4-speed gearbox operated by a car-type gear-change lever and gate. The engine unit included a kick starter and dynamo. Chain drive to the rear wheel was used. The additional features of a fully-sprung frame, weather protection for both engine unit and rider, efficient front and rear internal-expanding brakes, and good riding characteristics, made this design one of the most progressive for its time and a forerunner of certain designs which were commercially successful many years later. The efficiency and lightness of the engine unit caused it to be used with success for installation in experimental light aeroplanes; the A.B.C. engine unit (Inv. 1953–326) was actually used for this purpose.

The F.N. concern of Belgium produced after the war an advanced and elaborate machine of this kind which included a car-type gearbox and clutch, and a 748 c.c. engine (Fig. 4), having mechanically-operated overhead-valves, wet-sump lubrication and shaft drive. This machine was very suitable for sidecar use because of its power combined with flexibility. After the departure, in 1926, of Paul Kelecom, who had been responsible for the long and successful line of four-cylinder F.N. machines, the firm changed its policy and abandoned the four-cylinder type in favour of the simpler, more economically-produced single-cylinder, side-valve machine. A 10 h.p. (1145 c.c. capacity) four-cylinder motorcycle, the American Militaire, appeared in 1915, which included in its specification a unit-construction engine and gearbox unit, shaft drive and spring frame; it also had the unusual addition of a reverse gear and artillery-type wheels. It was intended for heavy sidecar work, especially for military purposes.

Generally, the four-cylinder formula was now found to be too expensive to build in Europe and in view of the efficiency and reliability of the simpler single and twin-cylinder machines, unnecessarily complicated. It was only in still prosperous places like America, where the added power and weight of this type was of advantage in the long distances to be covered, that the type continued.

The now well-established 1168 c.c. capacity Henderson was produced for some years in improved forms. In addition there appeared in the same country the 600 c.c. Cleveland and the 1229 c.c. Ace; the latter was originally designed by W. G. Henderson and was, in 1927, eventually taken over by the Indian concern and added to their list of models as the Indian 'Ace'. These models were on established American lines, with heavy frames, side-valve engines and all-chain drives, and were used to a considerable extent for mobile police duties.

There was a certain amount of interest in the possibilities of the radial disposition of cylinders for use in cars and motorcycles soon after the war, brought about partly by the influence of aero-engine development and partly by the desire to decrease size and weight. The 309 c.c. three-cylinder static radial Redrup engine of 1920 was a considerably more practical design than the earlier rotaries. There was a likelihood that it would have some commercial future, but few were made. Two of these three-cylinder units were installed as a single unit in an experimental machine, which was probably the first six-cylinder motorcycle to have been made. A three-cylinder design of 870 c.c. capacity, named the Eta, incorporated a robust and practical engine set across a specially designed cradle frame, which was sprung front and rear on leaf springs, and a shaft transmitted the drive to the rear wheel through a gearbox. The five-cylinder 640 c.c. Megola radial engine, of Swiss origin, was incorporated in the front wheel of a special design of motorcycle of somewhat impractical layout. But, although the radial engine had real advantages, such as smooth torque and light weight, it never became a commercial success when applied to motorcycles.

In 1928 the German, F. Opel, produced a rocket-propelled motorcycle as an extension of the similar tests he had already made with rocket-propelled cars. High speeds were obtained with this machine, but it remained a novelty and showed no advantages which could recommend its commercial production.

2½ h.p. Pullin-Groom motorcycle: 1920

2¼ h.p. Levis motorcycle: 1916

2¼ h.p. Velocette motorcycle: 1913

2¼ h.p. Triumph motorcycle: 1914

Plate 10

4 h.p. Triumph motorcycle: 1917

4/5 h.p. Zenith-Gradua motorcycle: 1920

3 h.p. A.B.C. motorcycle: 1919

DEVELOPMENT OF TWO-STROKE ENGINES

The active development of the two-stroke engine which was proceeding in 1914 was continued in this post-war period for the reasons of economy and the search for new and advanced types. The Scott design (Plate 11) continued to be produced. The new two-stroke motorcycles were of three main categories: a multitude of simple and cheap designs which, for the most part, adopted proprietary engines such as the Villiers, Dalm, Union and T.D.C.; some simple yet individual designs made by established manufacturers such as Levis, Velocette, Triumph and Royal Enfield; and some new and progressive designs, such as the Beardmore-Precision, the trunk-piston Dunelt, the horizontally-opposed cylinder, common combustion-chamber Black Prince, and the vee-twin cylinder Stanger. The Sun engine (Fig. 5) had a disc inlet valve to improve volumetric efficiency.

Of the first kind, the Villiers (Inv. 1921–1) and the Union (Inv. 1920–14) were the most widely used. They were of the normal three-port type, of robust construction and capable of prolonged hard work with little attention other than periodic decarbonising. The 269 c.c. Villiers engine, in particular was substantially the same as it had been in 1914, except that it now incorporated a magneto, and later a lighting-coil, within its flywheel. These were important innovations which were to become accepted practice. The Villiers Company continued to develop their designs and were soon producing various single-cylinder types ranging from 147 c.c. to 350 c.c. capacities. A 350 c.c. twin-cylinder engine was also produced but was not commercially successful. By 1930 this Company was the most important firm of proprietary two-stroke engine manufacturers in England.

The individual designs made by established manufacturers for the most part differed little from the 1914 models. There were, however, several interesting attempts to achieve increased efficiency by means of originality. The Beardmore-Precision machine used an engine of 350 c.c. capacity in a pressed-steel frame, which was sprung at rear and front on leaf springs. The Dunelt engine (Fig. 6), made in 500 c.c. and 250 c.c. capacities, employed a trunk-piston, working in a cylinder having two concentric bores of different diameters, for the purpose of improving charging volumetric efficiency. This was achieved by using the large diameter bore for the pumping action in the crankcase, and adding to it the contents of the annular space above the larger bore for transfer to the smaller or working bore. The Black Prince engine was a horizontally-opposed twin-cylinder engine, with a common crankcase and a common combustion chamber formed by joining the two cylinder heads by a large external finned pipe. The vee-twin cylinder Stanger engine employed two separate, but closely adjacent, crankcases; in addition, the angle of the two crank-throws was adjusted in relation to the angle between the cylinders, so that even firing occurred every 180° of crank rotation.

The $2\frac{1}{4}$ h.p. Pullin-Groom (Inv. 1931–40) (Plate 9) two-stroke motorcycle of 1920 was a sound attempt to produce a lightweight machine on novel and progressive lines. The single-cylinder engine, mounted horizontally in a frame consisting of two pressed-steel plates welded together, employed a mixing valve instead of an orthodox carburettor, and automatic poppet-valves for the admission of the fresh gas. A flywheel magneto, a spring frame and effective weather proofing were also features of this advanced design.

In spite of these novelties, it was the simple three-port engine, mounted in the standard frame, which continued to develop technically and be used consistently in commercial production, and the unusual and more complicated designs remained for the most part as curiosities.

This activity in two-stroke engine development also proceeded on the Continent, particularly in Germany. It was now that the German D.K.W., Schliah and Zündapp designs, and the Austrian Püch designs appeared, and so laid the basis for the considerable advances in two-stroke engine progress which were achieved by these concerns during the next twenty years. In particular, the deflectorless-piston loop-scavenge system of Dr. Schnuerle, which was adopted by D.K.W., the separate charging-pump racing engines of D.K.W., and the twin differential-piston design of Püch, were important advances. In the Schnuerle system, the inlet ports were so located that the gas stream was directed away from the exhaust ports to the opposite cylinder wall so that it rose up the cylinder and across the combustion chamber, thus assisting scavenging and avoiding admixture of the fresh gas with the residual exhaust gas. The twin-piston system adopted by Püch had the advantages of differential port timing, arranged by the location of both big-ends on one crankpin, and a uniflow motion of gas charging and scavenging.

MOTOR SCOOTERS AND MONOCARS

A more elaborate form of economical motorcycling than the cyclemotor unit made its appearance about this time in the form of the motor scooter and the more elaborate two-wheeled monocar. One or two isolated early forms of these vehicles had appeared before 1914, but after 1919 some of the designs were developed into practical vehicles which foreshadowed the wide development and use of this general type which began after 1945. Although these light, low-powered runabout vehicles were capable of a speed of more than 20 m.p.h., and of covering about 150 miles on a gallon of fuel, they were little used because of the relative cheapness of more elaborate and orthodox forms of motor transport at this period.

One of the simplest forms of scooter incorporated a Wall Autowheel as the rear wheel, and consisted of a simple open form of frame that was suitable for riders of both sexes. The Kenilworth was more elaborate, having installed in its main frame structure a 142 c.c. o.h.v. four-stroke engine which drove the rear wheel through a chain and belt. The most advanced of the scooters of this period was the A.B.C. Skootamota (Inv. 1948–202). It had a light open frame and small diameter wheels. A $1\frac{1}{4}$ h.p. high efficiency overhead-valve engine was mounted on a bracket over the rear wheel which it drove by means of a chain. The Autoped, which provided standing room only, used a small electric motor which drive the front wheel and was supplied from a large-capacity battery located on the frame. The Harper tricycle runabout of 1928 was intended to provide additional stability.

A form of monocar of considerable merit which was produced and used successfully for some years was the American Ner-a-car (Inv. 1934–499), designed by C. A. Neracher in 1919. It was a low open-frame motor bicycle having car features such as pivot steering and a bucket-seat for the rider. It had a 285 c.c. two-stroke engine which drove the rear wheel through a friction disc and a secondary chain; the friction disc was movable across the face of the flywheel, and so provided a variety of gear ratios. Later versions were fitted with more powerful Blackburne four-stroke engines.

A more elaborate and practical design which, since it combined good performance, comfortable travel and adequate weather-protection, may be classified as a monocar. This was the Ro-Monocar (Inv. 1953–382) which was designed, built and used in 1926 by the aviation pioneer, Sir Alliott Verdon-Roe. A 250 c.c. Villiers two-stroke engine drove the rear wheel through a plate clutch, 3-speed gearbox and a shaft and worm-drive gear. The driver was practically enclosed, and the upholstered bucket-seat and normal sitting position permitted by the design ensured a high degree of comfort for this class of machine.

ENGINE ACCESSORIES

Although there were considerable advances made in certain directions in the development of the accessories and ancillaries required for the motorcycle, the variety of them in this period declined and certain forms and types which were now proving to be the most suitable for general use were standardised. This simplification resulted from the actual operation of a wide variety of types during the first twenty years of the century. Apart from the prototype development of engines, and such accessories as gearboxes, drives, lighting systems, frames and suspensions, the 1920's saw the end of the specialised hand-made process of manufacture, and the gradual adoption of the mass-production method for motorcycles.

The development of magnetos had virtually become standardised by this time, having been established by the reliable Bosch design of 1914. English-built magnetos of the highest quality were now being produced by such concerns as B.T.H., Thomson-Bennett, E.I.C., Runbaken and C.A.V. The use at this time of a rotating permanent magnet and stationary primary and secondary windings was an added contribution to durability. Although this form gained acceptance in other fields, such as aero-engines, it was not generally adopted for motorcycles, which continued to use the old, well-established form of rotating wound armature and stationary permanent-magnets. The combination of magneto and electric lighting dynamo in one unit began after 1918.

The chief established designs of semi-automatic carburettors, produced by such manufacturers as Brown & Barlow, A.M.A.C. and Binks, continued to be made with general improvements, and it was with these instruments that the large proportion of production motorcycles were equipped.

In addition, a considerable number of new carburettor designs, most of which operated on the automatic principle, appeared after 1919. Few of these instruments survived, but they were in keeping with the originality of motorcycle design generally that was now being manifested. Because these new automatic instruments were not too accurate and dependable in their performances, the preference for the well-tried semi-automatic two-lever instrument remained.

The chief of these new automatic instruments were as follows. The Longuemare-Hardy (Inv. 1919–478) was constructed on car lines, with main, compensating and slow-running jets of the submerged type. The Claudel-Hobson, as fitted to the A.B.C. motorcycle (Inv. 1953–325), also employed the same principles as the type provided for cars and aeroplanes, with concentric jet tubes supplied with interacting bleed and compensating-holes. The Degory (Inv. 1920–781) had no jet in the ordinary sense, but vaporised the petrol in a small vacuum chamber before passing it to the choke-tube through a series of small holes; a slow-running jet was also used. The Cox-Atmos (Inv. 1927–1026), designed by A. Cox in 1918, had a jet which could be initially varied by a hand-controlled needle; two sleeve throttles, interconnected with one another and having cut-away portions, maintained correct mixture for all speeds and loads. The relation between the two sleeves could be adjusted for fine tuning.

Of these new designs of automatic carburettors, the most important was the Villiers, designed by F. A. Mills. This instrument was fitted with a hand-controlled throttle slide, to which was attached a tapering needle which varied the jet aperture with the throttle opening. Varied mixture characteristics were obtained by the use of different degrees of taper for the needle, and additional adjustment was provided by a screw which raised or lowered the needle in relation to the throttle-slide. This design proved very suitable for use with two-stroke engines, because of the flexibility of mixture adjustment it was able to provide.

Endeavours were now made to render the motorcycle engine quieter in operation, chiefly by the adoption of large and more efficient silencers and tail pipes, designed to reduce exhaust noise without wasting power through excessive back-pressure. The elongated tubular form of silencer incorporated with the tail pipe was now adopted, sometimes in conjunction with the older form of expansion chamber mounted in front of the engine crankcase.

TRANSMISSIONS

Although a few of the lighter forms of machines in the 1920's still fitted countershaft gearboxes of the chain-cum-belt drive variety, almost all the medium and heavy types of machines now used gearboxes of the all-chain drive variety. They were usually of the 3-speed type, with heavy duty, hand-operated plate clutches, and there were also some forms of 4-speed types. The gears were at first largely hand-lever operated and kick-starters were included as standard, even on small machines, so that 'run-and-vault' starting finally ceased. The positive form of foot-change gear control, which subsequently became a standard feature, was introduced in 1925 with the Velocette gearbox. This arrangement permitted the function of gear-changing to be performed rapidly and precisely without the necessity of removing a hand from the handlebars. The large variety of other and earlier forms of gears ceased to be used.

As has been indicated in the last paragraph, all-chain drive used in conjunction with a countershaft gearbox became standard practice. Even the few existing shaft-drives, in particular the pioneer F.N. design, disappeared. The efficient oil-bath chain case continued to be used by a few makers, such as Sunbeam, but before 1930 its use was illogically abandoned at about the time when other items of weatherproofing were being improved.

MOTORCYCLE ACCESSORIES

The motorcycle form of diamond frame had already become standardised in the previous period and, with the exception of a few specialist designs, such as the Scott, this practice continued. Frames developed little, except to become heavier and more robust to accommodate increasing power-outputs and the consequent higher speeds and load-carrying capacities of the machines.

Increasing speeds and engine powers had by 1930, however, begun to cause frame-flexing and 'speed-wobble' troubles, which were virtually unknown before, and as a result frame-suspension and steering-design techniques had to be considerably improved, particularly in racing machines. The Webb girder-type front spring forks, which were introduced in the 1920's, made a useful contribution to improved springing combined with good steering characteristics under both touring and racing conditions.

In addition, weather protection became of greater importance, particularly because of the competition of the baby car, so that such items as leg-shields, engine-shields and broader, better valanced mudguards became standard items of motorcycle equipment.

Considerable progress was made in the development of lighting systems after 1918. The older acetylene form still continued to be used on the grounds of economy, but it was little more efficient than it had been formerly. The dissolved-acetylene system was also used in conjunction with the Fallolite lamp, incorporating an incandescent pastille which considerably intensified the light. This arrangement was the highest development to which the acetylene system attained, but after a few years of commercial use it was entirely replaced by dynamo-battery system.

The improved forms of dynamo-lighting systems, tentatively adopted in 1914, were now vigorously developed, so that by 1930 systems, consisting of an efficient engine-driven dynamo, a cut-out and a storage accumulator, capable of giving brilliant illumination for head, side and tail-lamps, with the minimum of trouble and attention, were established as standard.

The simplest generator form of electric lighting consisted in using the unabsorbed half-cycle of the primary current of magnetos fitted to single-cylinder engines. Of these, the B.T.H. 'Sparklight' system was perhaps the most effective in the 1920's. This system consisted of an attachment to the magneto contact breaker, by means of which the primary winding could be tapped during the unused half-cycle, a small storage-battery and head and tail-lamps. Since, however, only some 3 watts were available under these conditions, the lighting capacity was limited. Another design of this type was the Runbaken 'Magneto-Lite', which operated on much the same principle.

Another form of magneto-lighting was the addition of a separate lighting coil in the Villiers flywheel-magneto, which gave an alternating current and could not therefore be used directly for accumulator charging but had to be applied directly to the lamps. Rectifiers were later added to this system for converting the alternating current to direct current for accumulator charging.

There were three separate general classes of dynamo lighting: (i) simple permanent-magnet dynamos, with or without accumulators; (ii) dynamos, having magnets separately excited and used with accumulators, which required a separate drive from the engine; and (iii) combined ignition and lighting magneto-dynamos. Although the first two systems were used to some extent, chiefly because they provided a more convenient alternative to acetylene-lighting, they soon gave place to the efficient, self-contained magneto-dynamo type of generator. The dynamo was generally superimposed above the magneto and driven from it by gearing, thus avoiding the necessity of a separate engine drive and making the whole a self-contained unit. The complete system included a cut out, a large-capacity storage battery and head, side and tail lamps; the output from the dynamo was between 30 and 40 watts, thus ensuring adequate illumination. This convenient source of electrical supply also caused the electric horn to be adopted, and the old bulb type of horn, which had been used on motor vehicles in general since the beginning of the century, to disappear.

Examples of the magneto-dynamo form of electric generator were the Lucas 'Magdyno' (Inv. 1935–192), the B.T.H. 'Mag-generator', the M.L. 'Maglita' (Inv. 1922–216) and the C.A.V. 'Dynamag'. These operated at 6 volts; voltage regulation was effected by the 'third-brush' method, and the cut-out disconnected the battery from the dynamo when current was not being generated.

SIDECARS

The need for economy, together with the higher performance of motorcycles during the 1920's increased the use of sidecar combinations as family vehicles to a certain extent, although the arrival in 1922–1923 of the real ultra-light or baby car such as the Austin Seven and Citroën Seven, as distinct from the relatively crude cyclecar of 1914, introduced serious competition. The balance of attractions remained in favour of the ultra-light car. Better, lighter and more comfortable sidecars were produced for use with most sizes of motorcycle, from the largest 8 h.p. machines to some of the smallest $2\frac{1}{2}$ h.p. machines. In particular, the springing was improved and attention was paid to weatherproofing, so that the designs included water-tight hoods and effective windscreens. The tendency towards greater comfort brought about the gradual adoption of larger-section — up to 3 in. — tyres of the wired-on variety which, after 1930, replaced the older, smaller-section beaded-edge type.

SOCIAL AND SPORTING DEVELOPMENTS

The unstable economic conditions of this period, which culminated in the worldwide depression of 1929–1930, naturally affected the purchase and use of motorcycles, and this, in turn, influenced the design and quality of the machines which were available. The growing inability of the average person to purchase new machines forced the manufacturers to cheapen and simplify their types so that not only prices, but also quality were lowered. Minimum new prices were reduced to little more than £20; moreover, an unnecessarily wide range of models were introduced in the attempt to attract all classes of buyers; this policy, however, caused prices to be relatively higher than if a few good models had been produced in substantial quantities.

A sound development in economical passenger carrying was the adoption of an adequate and relatively safe pillion seat. Pillion passengers had been carried on motorcycles since about 1910, but there was no actual provision for this. A cushion strapped to the carrier permitted a passenger to be carried in discomfort; the male sat astride and the female sat side saddle, without the additional support of footrests. Apart from the obvious danger of the side-saddle position — and a great proportion of pillion-passengers before 1914 were female — the steering and controls of that period made the management of solo motorcycles with a pillion passenger difficult and uncomfortable. In the 1920's, however, the improved controls and riding characteristics of motorcycles, together with the general adoption of the astride position by pillion passengers and the provision of adequately-sprung and padded pillion seats, such as the Tansad, made the practice of pillion-passenger carrying relatively safe and comfortable.

Tyres improved in quality and became more reliable. The formerly adopted beaded-edge form now began to be replaced by the wired-on type of tyre.

The winning speeds of the Tourist Trophy Races, always an indication of technical progress, increased from nearly 52 m.p.h. in 1920 to more than 74 m.p.h. by 1930, and in the same year the world motorcycle speed record was raised to 150.736 m.p.h., and 106 miles was covered within the hour on a 500 c.c. machine. Unfailing reliability under strenuous conditions over any distance, which was now an inherent quality of the touring motorcycle, was being attained under continuous maximum output conditions of operation. By 1930 the motorcycle was able to provide in actual use most of the social and utility advantages of which it was potentially capable.

This period of the 1920's was marked by the almost unchallenged excellence of British progress in motorcycle design and production. By 1930, however, an increasing challenge from Germany and Italy introduced new designs which laid the basis for future parity with, and even superiority to, British designs.

CHAPTER 6
Interim Development: 1930-1945

During the 1930's, motorcycle design and manufacturing techniques continued to be developed, in spite of the facts that this period began in financial disaster, continued in economic and international instability and ended in world tragedy. Against this sombre background the motorcycle, now firmly based upon well-established principles and well-proved designs, progressed in both advanced production and new experimental forms, until this activity was abruptly terminated by war in 1939. As had happened in 1914, the production of motorcycles was thereupon severely restricted to a few simple and robust single-cylinder side-valve designs such as those made by Norton, Royal Enfield, B.S.A. and Matchless, and some 400,000 of these were produced for war purposes.

In spite of the new prototype development in this country before the war, the greater part of commercial production after 1930 still consisted of old-established types, and of these the single-cylinder machine was the most used. Consequently, when Continental manufacturers, in particular Germany and Italy, began to put their new and progressive designs into production, and at the same time evolve still more advanced designs for racing and competition work, it was apparent that a major challenge was being made to long-established British supremacy. This was presently confirmed in various international races, and particularly in the Tourist Trophy Races, when machines fitted with supercharged horizontally-opposed twin-cylinder, wide-angle vee-twin cylinder and even more elaborate in-line four-cylinder engines began to demonstrate that they were potentially more efficient than the standard British formula of the single-cylinder machine.

It should be noted that, after 1930, while technical progress and improved outputs and performances continued to be achieved with both standard and advanced designs, it was no longer necessary to consider the question of reliability, with which other periods were so seriously concerned, as a special problem because this quality had by now become inherent in the motorcycle.

IMPROVEMENTS IN STANDARD TYPES

That perennial of British motorcycle design, the side-valve, single-cylinder, four-stroke machine, continued to be used in varying sizes from 150 c.c. up to 600 c.c. capacities, in spite of the increased production and commercial use of other more advanced types. The standard compression-ratio was maintained at a moderate figure — not more than 5·5 to 1 — in order to preserve flexibility. Power output and, unfortunately, overall weight increased without causing this type to lose anything of its identity, and better performance was obtained by the adoption of various detail improvements. The cylinder head, which had hitherto been formed integrally with the barrel casting, was now made detachable to facilitate decarbonising and valve grinding operations; it was sometimes cast in aluminium alloy with increased finning to promote the better cooling which the higher specific power outputs required. In addition to the side-valve operating gear, valve-tappets, springs and stems were also

Plate 11

$3\frac{3}{4}$ h.p. Scott engine: 1919

$2\frac{3}{4}$ h.p. Barr & Stroud sleeve-valve engine: 1922

$3\frac{1}{2}$ h.p. Triumph-Ricardo engine: 1921

490 c.c. Norton C.S.1 camshaft engine: 1928

Plate 12

348 c.c. Douglas motorcycle: 1947

192 c.c. Velocette L.E. 200 motorcycle: 1953

Douglas motorcycle: 1928

487 c.c. Sunbeam motorcycle: 1950

597 c.c. Ariel 'Square-Four' engine: 1934

650 c.c. Triumph vertical twin engine: 1935

350 c.c. Ehrlich engine: 1946

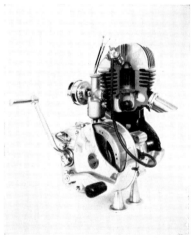

197 c.c. Villiers MK. 6E engine unit: 1952

Plate 14

192 c.c. Velocette L.E. 200 engine unit: 1952

494 c.c. Gilera 4-cylinder engine unit: 1953

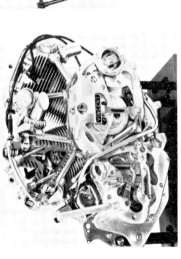

998 c.c. Vincent 'Rapide' engine unit: 1951

123 c.c. Lambretta motor-scooter engine unit: 1953

487 c.c. Sunbeam S. 7 engine unit: 1948

247 c.c. Alder engine unit and transmission: 1954

completely enclosed within a compartment formed with the cylinder casting, which protected the parts from dirt and permitted them to be continually lubricated by the oil spray from the crankcase. The lubrication system, hitherto generally of the 'total-loss' type, was now of the automatic mechanically-operated pump and dry-sump type; and the aluminium-alloy pistons were fitted with scraper-rings, so that high oil consumption and the consequent formation of carbon — both failings of earlier motorcycles — were effectively reduced. The decarbonisation period was increased from the former period of every 1000 miles to between 5000 and 10,000 miles. In addition, more durable valve steels provided greater reliability and longer life which, in turn, made the necessity for regrinding correspondingly less frequent. The side-valve engine became generally neater and more compact, and some designs even included unit-construction gearboxes. Perhaps the most typical of this long-established type at this time was the 596 c.c. 'Big-Four' Norton which, while retaining many of its original basic features, continued to be one of the most technically advanced and commercially successful of this type.

The overhead-valve type of single-cylinder engined machine improved considerably in detail design and was used to a considerable extent in push-rod operated form for commercial use and, in overhead-camshaft form, of both the single and double-camshaft types, for racing and high-speed work. Of the latter type, the Norton, Velocette and A.J.S. designs were particularly successful during this period. The Norton camshaft models of 350 c.c. and 500 c.c. capacities perhaps did most to maintain British prestige with repeated successes in international races; the establishment of a record lap during the 1938 Senior Tourist Trophy Race at an average speed of 91 m.p.h. was an outstanding example of its capabilities. The 1939 500 c.c. engine developed about 48 b.h.p. at a speed of 6500 r.p.m., operating on 80 octane fuel at a compression-ratio of 7·5 to 1. Not the least remarkable quality of such engines under sustained high outputs and performances was their almost unfailing reliability, which gave material substance to the claim that 'racing improves the breed'. It should be noted that this superiority was maintained with the basically simple and theoretically obsolescent single-cylinder engine against increasing Continental competition. Foreign firms tended to employ more complex and potentially more powerful multi-cylinder engines; these, however, had not yet attained the necessary reliability to use their greater outputs to full advantage. The most advanced of these foreign designs were perhaps the 250 c.c. and 350 c.c. single-cylinder Guzzi designs, the former of which won the Lightweight Tourist Trophy Races in 1935 and 1937.

The long-established vee-twin cylinder engine also continued to be used to some extent, but this type now either tended to become larger in side-valve form, such as the 1140 c.c. Royal Enfield, or to adopt overhead-valve gear in the smaller sizes, such as the 500 c.c. B.S.A. and the New Imperial designs. Some special models, such as the Vincent-H.R.D., the MacEvoy and the Brough 'Superior' S.S. 100, some of which were fitted with superchargers, were used for high-speed record attempts. In 1937 a Brough Superior motor-cycle, fitted with a 1000 c.c. J.A.P. engine, established the world speed record at nearly 170 m.p.h.

The horizontally-opposed twin-cylinder engine had been adopted in the 1920's by the German B.M.W. firm, and it was during this period that they began to develop it in highly specialised shaft-driven forms for both racing and high-speed sporting uses.

The smooth running, efficient cooling and high power output of which this form of engine was capable, combined with the ease with which it could be adapted to the compact unit form of construction, combined to maintain it as an effective layout. Moreover, not only did it prove itself to be capable of high powers when normally aspirated, but it also showed a high degree of reliability as a supercharged unit, as was demonstrated when a 493 c.c. B.M.W. machine fitted with a supercharger won the 1939 Senior Tourist Trophy Race at a record average speed of 89·38 m.p.h., and a 493 c.c. supercharged B.M.W. machine in 1937 increased the world speed record to more than 173 m.p.h.

The shaft-driven horizontally-opposed cylinder arrangement set across the frame was also tested in prototype form by the Douglas Company when, in 1935, their 'Endeavour' model appeared. This interesting design departed radically from long-established Douglas practice by the adoption of a 500 c.c. horizontally-opposed side-valve engine and gearbox unit, which drove the rear wheel through a shaft and bevel gear. The cylinders were arranged transversely within a wide cradle frame, which provided protection for the cylinders as well as a stiff structure.

NEW FOUR-STROKE ENGINES

Apart from these standard and well-established commercial forms which provided the bulk of production machines, this period was chiefly notable for the appearance of certain new engine forms, some of which were eventually widely adopted in commercial production.

The chief of these new forms was the vertical twin-cylinder engine which, after 1945, became one of the most popular types. This cylinder arrangement provides the smoothness of the twin-cylinder arrangement, with the small and compact bulk of the single-cylinder engine and, mounted across the frame, even cooling of the two cylinders is provided. On the other hand, its balance with a 360° crankshaft is no better than a single-cylinder engine. The 650 c.c. Triumph Twin engine (Inv. 1935–52) (Plate 13) was one of the first examples of this type which went into large commercial production. The pistons of this design moved together and therefore permitted even firing periods; this arrangement became standardised in preference to the earlier arrangement of cranks set at 180° to each other to obtain better balance characteristics at the expense of uneven firing periods. Overhead valves were generally adopted in the interest of high efficiency, although side-valve versions were adopted for military use to simplify maintenance.

Another form of unorthodox twin-cylinder engine, which first appeared in 1929, was the 400 c.c. Matchless 'Silver Arrow' model. This design was a narrow-angle (26°) twin-cylinder unit set longitudinally in a cradle frame, and having the two cylinders formed in a single casting. The importance of this design was that it foreshadowed a more elaborate one of greater promise, namely, the 593 c.c. Matchless 'Silver Hawk' engine (Inv. 1935-31), mentioned later.

An original form of vee-twin cylinder engine was introduced by the Italian Guzzi factory. This firm had, since 1930, been producing high output, overhead-valve, single-cylinder engines; the 1935 and 1937 Lightweight Tourist Trophy Races were won with 248 c.c. examples of this make at the respective average speeds of 71·56 m.p.h. and 74·72 m.p.h. The 499 c.c. vee-twin cylinder engine had its cylinders disposed at 120° to each other, an arrangement which permitted improved cooling and also provided a lower centre of gravity when mounted in the motorcycle. This machine was the first to challenge a long-established British superiority in the Tourist Trophy Races by winning the Senior event in 1935 at an average speed of 84·68 m.p.h.

The 593 c.c. Matchless 'Silver Hawk' engine (Inv. 1935-31) had four cylinders arranged in narrow-angle (18°) pairs formed within a monobloc casting, with a detachable cylinder head incorporating an overhead-camshaft and valves. This layout constitutes the most compact four-cylinder arrangement possible, and it is therefore very suitable for motorcycle use; moreover, with its short, stiff crankshaft and compact arrangement of connecting rods, the type is efficient and reliable, as has been consistently demonstrated since 1922 by the similar Lancia car engines. This original Matchless design was in production for only a short time, but it is possible that the type will be revived to serve the modern trend towards four-cylinder motorcycles. A more complicated, and at the same time more commercially successful, four-cylinder engine, was the 597 c.c. Ariel 'Square Four' (Fig. 7) design (Inv. 1935-68) (Plate 13), which first appeared in 1929. The four parallel cylinder bores were arranged equidistant from one another in an air-cooled monobloc casting, and the detachable cylinder head carried the overhead-valve gear. This arrangement was more bulky and heavier than the narrow vee four-cylinder unit already described, since two separate geared crankshafts were used. Each pair of crankpins were set at 180° to each other, and the four firing periods were evenly spaced through two revolutions. This type has been in production since 1929 but, although the Ariel Company made a commercial success of it, the type was not then adopted by any other maker. The engine capacity of later models was eventually increased to 997 c.c., in which form the Ariel 'Square Four' model made a reputation for smooth, high-speed travel, both as a solo as well as a sidecar machine.

The 1265 c.c. Indian 'Ace' was the best known of the few large four-cylinder machines with in-line engines produced in this period, although occasionally other designs appeared for a short while, such as the Danish 746 c.c. Nimbus. The Nimbus had an air-cooled in-line car-type overhead-valve engine unit, with clutch and gearbox and final shaft-drive, which was mounted in an open pressed-steel cradle frame of light and simple design. The whole machine weighed 380 lb., which was low for this elaborate type.

Of more significance than these touring designs was the appearance from the Continent of the high performance four-cylinder machine which was originally intended for racing, and later gave promise as a future production motorcycle engine. Although more complicated and expensive to make, the four-cylinder engine with its smooth torque, low transmission loading, and high potential powers and speeds had much to recommend it. The first example of this high performance type of four-cylinder engine was the 494 c.c. Gilera, which covered more than 121 miles in one hour in 1938 and showed remarkable speed and acceleration qualities in various international events in the following year. This design, with its first successes in international events, demonstrated the considerable potentialities of this type.

ROTARY-VALVE ENGINES

There have been several attempts to apply the rotary valve in various forms to the internal combustion engine and, in the motorcycle field, include such designs as the Corah (1910), the Scott (1913) and the Brough (1914). Two of the most successful proposals were made in four-stroke form during the 1930's by R. C. Cross and F. M. Aspin respectively. The Cross design employed a barrel form of valve, and elaborate arrangements were made for adequate lubrication and the maintenance of correct clearances under the pressure and temperature differences that occur during operation. Various designs of from 250 c.c. to 500 c.c. capacities were produced for motorcycle use, including a special 500 c.c. engine for the 1935 Senior Tourist Trophy Race.

The Aspin rotary-valve was of conical shape and actually formed the combustion head of the engine. One of the chief features of this design was that the sparking plug was masked by the valve until the ignition point was reached, and there was therefore no tendency to pre-ignition from hot plug points. This feature, combined with the simple and well-shaped combustion head formed by the conical valve, enabled the claim to be made that compression ratios as high as 10 to 1 could be used with fuels of as low a grade as 60 octane rating, with resulting high thermal efficiencies. In spite, however, of these and other efforts to perfect different valve forms for the petrol engine, the poppet valve continued to remain unchallenged for use with four-stroke engines.

NEW TWO-STROKE ENGINES

Although a few of the original two-stroke engine manufacturers, notably the Scott Company, continued to use the deflector-type piston and normal port arrangement in two-stroke engine designs, the 1930's are noteworthy for the substantial adoption of the Schnuerle flat-top piston and directive-port technique. This new and more efficient arrangement of the difficult problem of ensuring the most efficient shape and location of ports had first been adopted in Germany by the D.K.W. concern with considerable success, and later by Zündapp, Victoria and others. The Villiers Company, which by now had gained a virtual monopoly of proprietary two-stroke motorcycle and small industrial engines in Great Britain, soon brought out their own version of this arrangement, which consisted in providing two pairs of impinging gas streams directed up the cylinder bore, with two opposed exhaust ports located between these pairs of inlet ports. The improved efficiency and greater powers obtainable with the Schnuerle arrangement confirmed that it more nearly conformed to the true gas flow in a two-stroke engine cylinder than did the older deflector-piston arrangement.

The range of Villiers two-stroke engine designs was comprehensive and included different capacities from 98 c.c. to 350 c.c., which were used for a variety of vehicles ranging from autocycles to three-wheeled runabout cars.

The later designs included flat-topped pistons and the Villiers form of port arrangement. In 1935 98 c.c. and 125 c.c. engines of unit construction were produced which included a 3-speed gearbox and clutch. This type of unit eventually became standard practice.

An example of the later large-capacity two-stroke motorcycle engine is seen in the 560 c.c. engine (Inv. 1952–315) of 1934. Although essentially simple in general design, it was capable of sustained high output because of the use of a well-finned aluminium-alloy barrel with an austenitic cast-iron liner. The similar expansion coefficients of these two metals ensured constant contact between them, and so provided uninterrupted heat flow and efficient cooling.

The provision of reliable and relatively powerful two-stroke engines of less than 100 c.c. capacity brought into being an ultra-light and simple form of motorcycle which, fitted with the minimum of equipment such as a 2-speed gearbox, sold at a low figure and served as a handy runabout for short journeys. Although this form declined by about 1936 for a time, to give place to the autocycle, it served as the forerunner of improved and more powerful designs of similar capacity which, after 1945, became capable of long-distance touring. The autocycle was also intended as a low-powered economical runabout, but included some cycle features in its design, such as pedalling gear, a high saddle position and an open frame to make it suitable for both sexes. The pedalling gear was intended to make a 2-speed gear unnecessary and also to add to the ease of starting and improve handiness in traffic, but some later versions eventually included both gears and clutch, while retaining the pedalling gear.

The Scott design of two-stroke motorcycle continued to be made generally on its original lines, with periodic improvements. In 1935, however, a new and more elaborate Scott design was produced in prototype form which included a 750 c.c. three-cylinder in-line engine of advanced characteristics and high performance. This water-cooled unit had an elaborate built-up crankshaft, which ran on ball and roller bearings and permitted the use of separate crankcase compartments necessary for the three-port two-stroke form of operation. A clutch and 3-speed gearbox were combined with the engine, thus making the whole a self-contained unit. Economic rather than technical difficulties prevented this interesting design from going into commercial production, not only for motorcycles but also for light cars.

The greatest advance in two-stroke motorcycle engine design at this time, however, was made on the Continent and particularly in Germany which had, in 1929, originated the efficient Schnuerle loop-scavenger system. The D.K.W. firm produced a range of machines including single-cylinder models of 248 c.c. and 350 c.c. capacities, and a twin-cylinder model of 500 c.c. capacity; these were made in large numbers and used extensively in Europe, and to some extent in this country. In addition, special high-powered designs were also developed for racing, including a twin-piston arrangement permitting differential port timing, in conjunction with a separate pump, which could be arranged to provide either atmospheric or supercharged fresh-gas charging. In the latter form, the 248 c.c. version was very successful in the years immediately before 1940; with it was won, in particular, the 1938 Lightweight

Tourist Trophy Race at an average speed of 78·48 m.p.h. The Austrian twin-piston Püch design in twin and four-cylinder forms, together with other advanced types, such as the Zündapp, Victoria and Imperia of German origin, were increasingly successful and provided efficient alternatives to four-stroke engined designs for Continental users.

The original and progressive development and wide use of the two-stroke motorcycle on the Continent at this time prepared the way for still further important improvements and wider use in the years after 1945.

ACCESSORIES

Since the basic necessities in respect of motorcycle accessories had been evolved in the 1920's, this later period contributed detail improvements rather than new developments. Some experiments were made with pressed-steel frame construction, which made possible lighter, stronger and cheaper frames; for the most part however, these new forms remained experimental, and the tube-and-socket method continued to predominate, although the construction became heavier to provide for the higher outputs and faster speeds. By 1938 the average weight of a 500 c.c. machine had risen to more than 400 lb, which was about twice the weight a machine of similar engine capacity had been twenty years before. The all-enclosed engine form of machine was also tried as a means of giving the motorcycle a smooth, weather-proof and easily-cleaned exterior; the Coventry Eagle 'Pullman' and the Francis Barnett 'Cruiser' were examples of this type which were in production for some years. This technique was even developed to the extent of using completely-enclosed streamlined machines, such as the B.M.W. and Gilera, for maximum speed attempts.

Twist-grip throttle controls and foot-operated gear change mechanisms were standardised soon after 1930.

Brakes tended to become larger and more efficient, to deal with the increased speeds and weights of motorcycles. Almost without exception, they were of the internal-expanding type and some were air-cooled for racing purposes. There were a few examples of frame-springing; the O.E.C. machine (Inv. 1936–454), for instance, incorporated rear-wheel springing in a duplex tubular frame. There was, however, as yet no real tendency to adopt this effective yet somewhat expensive method of improving riding comfort. Spring forks generally remained the same with the improved Druid, Brampton and Webb designs in considerable use, but progress was indicated by the adoption in 1938 by B.M.W. of the tele-hydraulic type, foreshadowed before 1914 by such early telescopic types as the Scott, F.N. and Chater-Lea forms. Another original form of front fork was the O.E.C. (Inv. 1936–454) which employed an original link form of steering arrangement. Soon after 1930 the beaded-edge type of tyre was entirely replaced by improved tyres of the wired-on variety.

The old form of rear stand, which was not too easy to operate since it required the rear portion of the now quite heavy machine to be lifted, began to be replaced by the simpler and easier lean-to or pull-up forms of stand. The electric system was improved by the adoption of the constant-voltage type of dynamo in place of the older form of 'third-brush' voltage control. Dipping switches to minimise headlight glare were also standardised.

SOCIAL AND SPORTING DEVELOPMENTS

By 1939 the motorcycle had achieved its full impact upon social life. It was accepted in Great Britain and the Continent as a practical and economical form of personal transport, despite severe competition from the really economical and reliable miniature car, such as the Austin Seven. In the U.S.A., on the other hand, the importance of the motorcycle had seriously declined, because of the general adoption of the large motorcar.

Motorcycle events of all kinds were increasingly popular, particularly among younger users, and it is to be noted that this enthusiasm derived solely from sporting and technical interests, and not from motives associated with betting. Such events included road and track racing, world speed-record attempts, road-touring trials of one to six days duration, cross-country 'scrambles' or 'moto-cross' as they were known on the Continent, and even speedway racing. The Brooklands race track, which had contributed to motorcycle, motorcar and aeroplane development for some thirty years, was closed at the end of 1939 and adapted for the uses of aircraft production.

CHAPTER 7
Later Development: 1946-1955

The year 1945 ended, as had the year 1919, a gap in the continuity of motor-cycle design and construction, during which the new designs of 1939 were not produced and the still newer and more progressive designs, which would normally have appeared during this period, were not evolved. Progress in general technique, however, had continued, again particularly in the aero-engine field, and considerable new knowledge of a general kind was available for the production of the post-war designs.

This progress was, however, made in the face of extreme national and international financial stringency, which necessitated the exporting of as many of the new production machines as possible; in addition, there was a world shortage both of materials and petrol. Motorcycling was, therefore, severely restricted for some years, particularly in Great Britain, and this factor, coupled with the very sharp rise in prices of up to three times the 1939 figures, brought about both here and on the Continent a move towards strict economy. This situation quickly produced and popularised large numbers of relatively cheap and low-powered machines of the cyclemotor unit, the autocycle, the ultra-lightweight motorcycle, and the motor scooter and monocar varieties.

Still further improvements were achieved with diminutive engines of from 18 c.c. to 98 c.c. capacities when higher-grade fuels once more became available; higher compression ratios and more efficient combustion chamber design were used to increase specific power output and decrease fuel consumption. After 1950, these small-capacity machines consequently became sufficiently powerful and robust to be established as practical vehicles for both solo and passenger-carrying purposes. Although the larger and more orthodox types of motorcycles continued to be used in considerable numbers, the perfecting of the smallest capacity form of machine induced a large number of people, who otherwise would not have done so, to adopt this form of motorcycling.

This post-war period was remarkable for the considerable increase in specific outputs of small four and two-stroke engines, and for the large variety of motorcycle types in production. Some two-stroke designs of only 98 c.c. capacity, for instance, were capable of developing nearly 4 h.p. at speeds of over 8000 r.p.m., with a high degree of reliability, and four-stroke engines of similar size and equipped with overhead-valves were equally powerful. This substantial increase of specific output had the important effect of reducing the engine sizes necessary for various purposes, so that the former average of 500 c.c. capacity for a practical touring machine began to be replaced by capacities of 200 c.c. and less.

Types ranged from such machines as the 120 m.p.h. 998 c.c. Vincent twin-cylinder machine (Plate 14) which, in its most expensive form, cost £474; through the 650–500 c.c. single and twin-cylinder machines, costing from £200 to £250; and the 350 c.c. and 250 c.c. models of good performance, costing £100 to £180; to the 200 c.c., 125 c.c. and 98 c.c. lightweights, mostly two-strokes, costing in the region of £120; and, finally, the cyclemotor units,

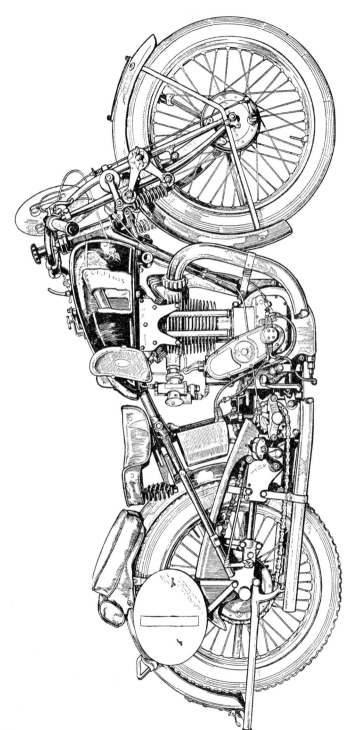

Fig. 1 493 c.c. Isle of Man Sunbeam motorcycle: 1929

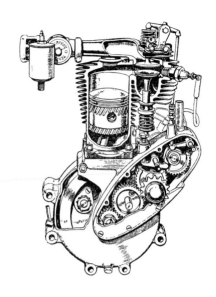

Fig. 2   499 c.c. Rudge-Whitworth engine: 1921

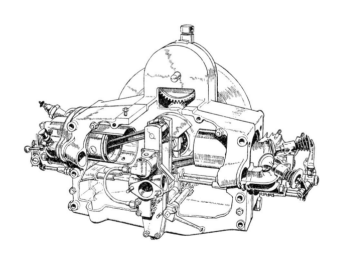

Fig. 3   500 c.c. oil-cooled Bradshaw engine: 1922

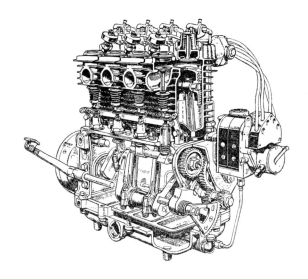

Fig. 4 748 c.c. four-cylinder F.N. engine: 1923

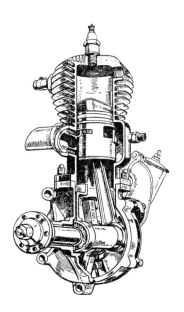

Fig. 5 2¼ h.p. Sun two-stroke engine: 1922

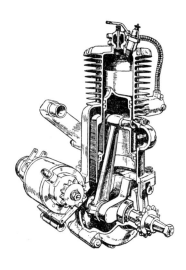

Fig. 6   500 c.c. Dunelt two-stroke engine: 1922

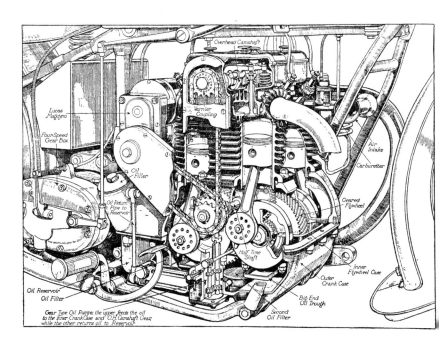

Fig. 7   597 c.c. Ariel 'Square Four' engine: 1935

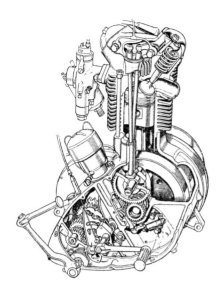

Fig. 8　149 c.c. Triumph 'Terrier' engine and unit gearbox: 1954

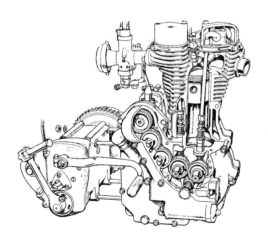

Fig. 9　349 c.c. Royal Enfield engine and gearbox: 1953

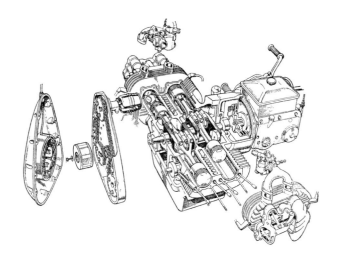

Fig. 10 498 c.c. Wooler 'Flat-Four' engine and gearbox: 1953

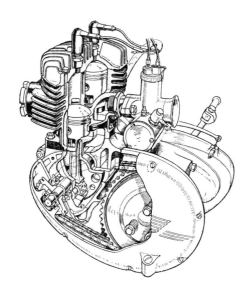

Fig. 11 322 c.c. Anzani two-stroke twin-cylinder engine: 1954

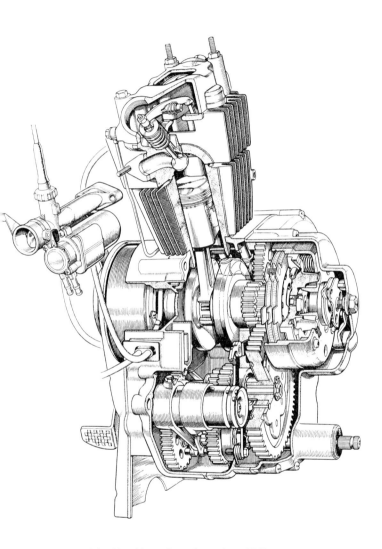

Fig. 12   49 c.c. Honda engine: 1962

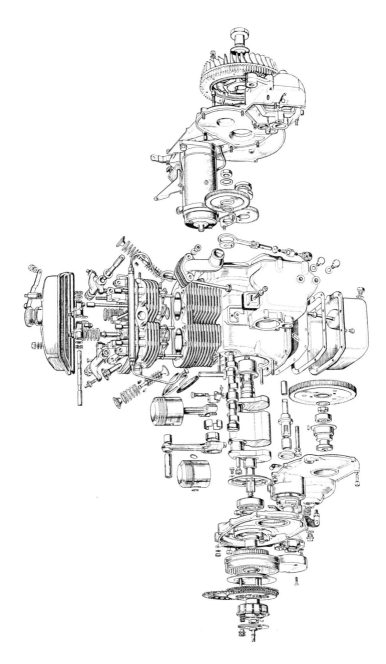

Fig. 13   249 c.c. B.S.A. Sunbeam motor scooter engine: 1958

costing about £20. In addition there were a number of motor scooters and monocar runabouts, mostly from the Continent, with prices ranging from £70 for the simpler types to £150 for the more elaborate forms.

With this great variety of motorcycle types being made available in increasing numbers, to serve as economical and satisfactory means of personal transport for one or two people as well as a considerable amount of baggage, the number of motorcyclists in Great Britain increased by 1953 to about a million. These users now included not only the youth of both sexes, but also a considerable number of more elderly people.

NEW VERSIONS OF ESTABLISHED TYPES

In spite of the variety of new and more complicated engine types which now began to appear in quantity production, the simple single-cylinder form continued in considerable popularity in Great Britain, and was produced in still more improved and efficient forms, for both touring and racing uses. This continued development was effectively demonstrated by the ability of this relatively simple design to give still higher outputs – by 1953 over 60 b.h.p. was being obtained from racing engines of 500 c.c. capacity – together with great reliability. These characteristics enabled the type, particularly as produced by the Norton firm, to maintain its superiority in arduous road races, even in the face of the growing challenge from the more elaborate and potentially faster multi-cylinder types. The Continental four-cylinder machines however, were now acquiring reliability as well as power and speed, and the year 1955 showed their superiority over the single-cylinder engine for racing purposes.

This reliability combined with high performance of the single-cylinder engine was applied effectively to the current production models of this type, for both solo and sidecar use in sizes from 650 c.c. to 500 c.c. capacities and for solo use in sizes from 500 c.c. to 150 c.c. capacities. Among the larger sizes were the Norton, B.S.A., Triumph, Royal Enfield (Fig. 9), Ariel and P. & M. designs; and the smaller sizes included the A.J.S., Velocette, B.S.A., and Triumph 'Tiger Cub' and 'Terrier' (Fig. 8), designs.

In addition the improved side-valve form of single-cylinder engine also continued to be used, although to a lessening degree, in the interests of reliability and flexibility of operation, simplicity of maintenance and low first cost. The chief of this type were the 'Big Four' Norton, B.S.A. and Ariel designs in the larger capacities, as well as some others in the smaller capacities.

A notable development after 1945 was the large production of the vertical twin-cylinder type of machine, in sizes of from 700 c.c. to 350 c.c. capacities, by several of the more important manufacturers, and the wide adoption of it by users. This type, which had been pioneered in commercial production by the Triumph Company in 1935, was developed vigorously by this concern so that their 'Speed-Twin' model (Inv. 1950–263) and their later and larger 'Thunderbird' and 'Tiger 110' models enjoyed a wide sale for both private and official purposes; the former model was adopted as the standard mount for mobile traffic policemen and, in a derated side-valve form, was adopted for Army despatch riders. In improved form, such as the 650 c.c. Tiger 110, with enlarged inlet ports, a larger bore carburettor, faster cam-timing and a compression ratio of 8·5 to 1 for use with fuels of the higher octane ratings, some 42 h.p. was produced at a crankshaft speed of 6500 r.p.m. Vertical twin-cylinder models of high power and efficiency were also produced in considerable quantities by such manufacturers as Norton (Fig. 23), B.S.A., Ariel, A.J.S. and Matchless (Fig. 24), and the type was adopted on the Continent and even in America. The overall layout of these individual designs was generally similar, incorporating vertical cylinders in monobloc form, with a detachable cylinder head and overhead valves operated by push rods, and a built-up 360° crankshaft which caused both pistons to move together.

A feature of aero-engine development which was now successfully applied in commercial form to motorcycle construction was the use of the aluminium-alloy barrel, having a cast-iron liner of substantially the same rate of expansion. Besides resulting in improved heat-flow characteristics, which provided adequate cooling for still higher specific power outputs, this form of construction also resulted in a useful saving in weight.

The large vee twin-cylinder form of engine of from 750 c.c. to 1200 c.c. capacities, which had remained an important standard type up to 1939, had now virtually disappeared except for the 998 c.c. Vincent design and a few American examples. The considerably higher specific outputs of engines now obtainable, however, made these large-capacity designs generally unnecessary. In addition the smaller sizes of this type were also entirely replaced by newer forms, in particular the compact vertical twin-cylinder type.

NEW FOUR-STROKE ENGINES

Progress after 1945 in four-stroke engine equipped machines tended in general to make use of such progressive features as high efficiency multi-cylinder camshaft-engines of unit construction, which drove the rear wheel through a shaft, and were mounted in sprung cradle frames designed to give the most stable as well as most comfortable riding characteristics possible. In addition, after 1950, there was a strong trend, particularly on the Continent, to develop the highly-efficient single-cylinder engine in sizes of from 60 c.c. to 200 c.c. capacities.

One of the most notable of the new designs for all-round utility was the Velocette L.E. 150 (Inv. 1952–49) of 1948. The engine and gearbox unit (Inv. 1952–50) was of compact construction, having a water cooled horizontally-opposed engine of 149 c.c. capacity, later increased to 192 c.c., set across a cradle frame and totally enclosed by integral shields which protected both engine unit and rider. The addition of a windscreen to the handlebars gave this machine a weather protection almost equal to that obtained with a monocar. The car-type gearbox and multi-plate clutch, together with the final shaft drive and exceptionally smooth running of the small well-balanced engine, combined to provide a new concept of easy, comfortable and trouble-free motorcycling. (Plate 14.)

Somewhat more conventional in frame design, but as original and progressive in respect of the engine unit, was the Sunbeam S.7 (Inv. 1951–193) model of 1950. This machine had a 487 c.c. twin-cylinder vertical engine, with cylinders and crankcase formed in one piece; a 4-speed integral gearbox, a single-plate clutch, an overhead camshaft, coil ignition, and a final shaft-and worm-gear drive, to the rear wheel were used The cylinder barrels were of aluminium alloy and had liners of austenitic iron, so that both barrels and liners expanded under heat at the same rate, thus ensuring an uninterrupted heat flow. The connecting rods were aluminium-alloy forgings with lead-bronze bearings, and an Amal carburettor and an air-cleaner were fitted. A pump supplied oil at high pressure to all bearings, and the engine unit was mounted in rubber on the bearers of the cradle-type frame. Riding comfort was improved by the use of telescopic front forks equipped with hydraulic dampers, and the rear of the frame had a plunger-spring suspension, in addition, large-section low-pressure typres (4·50 in. by 16 in.) were fitted to interchangeable wheels. The engine developed 25 b.h.p. at 5800 r.p.m., and the machine was suitable for both solo and sidecar use. (Plate 14.)

Another high performance twin-cylinder machine of unusual design, which was specially developed for racing, was the 500 c.c. A.J.S. The engine was a parallel twin-cylinder unit, mounted in a cradle-frame. The aluminium-alloy cylinders were inclined forward and both cylinder barrels and heads were equipped with deep cooling fins to increase the effective cooling area. Although of progressive design and capable of high performance, this model did not appear in production form, but it figured largely for a few years in racing events from 1959.

C. Wooler, a pioneer of motorcycle design who was noted for his novel projects, was responsible for two interesting and progressive four-cylinder designs during this period. The first engine of 1950 (Inv. 1952–301) depended for its action upon the motion of a single centrally-pivoted bell-crank, one end of which was in turn motivated by the sequence of impulses from the four short connecting rods, while the other end formed the small end for the single connecting rod, by means of which the single-throw crankshaft was rotated. This arrangement, however, had the disadvantage of creating high inertia-loading losses at high speeds, and so its maximum useful performance was limited. The form of construction, however, resulted in a light and compact unit which, mounted in a light cradle frame incorporating front and rear springing and final shaft drive, permitted the whole motorcycle to weigh less than 200 lb., and possess an exceptionally low weight-power ratio. The second design of 1953 (Fig. 10), known as the W.F.F.I. model, was a 498 c.c. horizontally-opposed four-cylinder machine; this was of conventional, although advanced type, with a two throw crankshaft, push-rod operated overhead valves and forged aluminium-alloy connecting rods. The engine, in commercial production form, was capable of developing 32 h.p. at 6000 r.p.m. The drive to the rear wheel was through a 4-speed gearbox, hand-operated plate clutch, and final shaft and bevel-gear drive with Hardy-Spicer universal joints. The weight was under 300 lb.

The most notable of the larger designs was the 998 c.c. Vincent, which was designed on weight-saving and good-handling lines, without sacrificing maximum output and performance, so that it could be used as a high performance solo as well as sidecar mount. One of its chief features was the use of the engine unit (Inv. 1951–1) as an integral part of the frame structure which, combined with the extensive use made of aluminium alloy in the construction of the engine, materially reduced the overall weight — 380 lb. — of the machine. The engine, as tuned for the 'Black Shadow' speed model, was capable of developing 55 b.h.p. at 5700 r.p.m., and the production model was capable of a speed of 120 m.p.h.

Considerable advance was made after 1950 in certain Continental four-cylinder engines and machines, so that they became an even more serious challenge to the simpler British single-cylinder machines, at least for racing if not for actual commercial production. The more important were the 494 c.c. Gilera (Plate 14), the M.V. and the Guzzi, all of Italian origin, and by 1953 such designs were capable of developing some 65 h.p. at crankshaft speeds approaching 8000 r.p.m. The two former had their engines set across the frame for the purposes of ensuring even cooling and also to provide a reasonably short wheel base. These multi-cylinder, high-efficiency designs were capable of very high speeds and rates of acceleration, together with excellent road-holding and steering characteristics, and they gave promise of becoming the most efficient form of road-racing motorcycle. Whether, however, their greater complication and cost of manufacture could permit them eventually to become suitable machines for commercial production and everyday use was, in 1954, still an open question.

The first British manufacturer to follow this Continental lead was the old established firm of J. A. Prestwich and Co. Ltd., which in 1953 produced a 500 c.c. four-cylinder water-cooled vertical prototype engine with twin overhead-camshafts.

There was also a great advance made after 1950 in the development of small four-stroke single-cylinder engines of from 60 c.c. to 200 c.c. capacities which, mounted in advanced machines using such features as light, single-member spine frames, front and rear-wheel springing and shaft drive, began to challenge the 500 c.c. and 350 c.c. capacity machines as practical touring mounts. Among these may be mentioned the 150 c.c. Triumph 'Terrier' (Fig. 8), the 60 c.c. and 98 c.c. Ducati, the 150 c.c. Aero Caproni Trento, and the N.S.U. 98 c.c. 'Fox' and 247 c.c. 'Max' designs. Some of these small engines were capable of power outputs of the order of 70 h.p. per litre capacity at crankshaft speeds up to 6000 r.p.m. The 247 c.c. Adler engine unit and transmission (Plate 14) illustrates the trend towards compactness of design.

Of the small horizontally-opposed form of engine, the 250 c.c. Zündapp design, which developed $18\frac{1}{2}$ h.p. at 7000 r.p.m., was of considerable merit.

GAS-TURBINE ENGINES

The instinct for progress during this period even caused the possibilities of the gas-turbine-engined motorcycle to be considered, in view of the successful establishment in commercial use of this form of prime mover since 1945. In 1953 it was considered that the gas-turbine engine in a motorcycle would be even lighter and of higher specific power output than the normal piston engine and gearbox; in addition, starting, acceleration and smoothness of operation would all be superior, but it would be less thermally efficient. It would, therefore, have a higher fuel consumption, and its unorthodox shape would necessitate basic modifications in frame design.

Further important progress in the design of small two-stroke engines was made after 1945 in various countries and, as a result, a variety of designs of capacities up to 250 c.c. were developed and put into large production. These were adopted for use in the many forms of lightweight vehicles which, for reasons of economy and convenience, had now become popular. Detail improvements in port design involving both location and shape to promote smoother and more efficient gas flow, together with increased compression ratios and maximum crankshaft speeds, raised specific power outputs by considerable amounts and also decreased fuel consumption. Also, with improved materials and manufacturing techniques, which included such items as hard-chromium-plated aluminium-alloy cylinder bores, cast-iron liners permanently bonded to aluminium-alloy cylinder barrels, and aluminium-alloy pistons incorporating inset steel piston-ring carriers, the durability of the two-stroke engine was considerably increased, so that about 30,000 miles could now be covered before any major engine overhaul became necessary.

In England the simpler single-cylinder forms were largely produced by the Villiers Company, and one of the most widely used was their 197 c.c. Mark 6E engine and 3-speed gear unit (Inv. 1952–301) (Plate 13), which was capable of developing 8·4 h.p. at 4000 r.p.m. This unit was used for the new form of cyclecar runabouts, such as the Bond and the Pashley, as well as for numerous motorcycle designs. A larger version of this type, the 225 c.c. Model 1J engine and 4-speed gear unit, developed 10 h.p. at 4500 r.p.m., and embodied such refinements as the enclosure of all parts within a smooth streamlined shell. At the lower end of the range, the 98 c.c. MK.4F., which was much used for a great variety of purposes, developed 3 h.p. at 4000 r.p.m., which was one-third higher than the output of the 1939 MK.2F. of the same basic design. Moreover, this diminutive engine was capable of long-distance touring at average speeds of as high as 40 m.p.h.

The 224 c.c. Excelsior 'Talisman' had a vertical twin-cylinder engine set across the frame. A larger version (422 c.c.) (Fig. 11) of the same general arrangement was produced by the British Anzani Engineering Company in 1953, which had the added advantage of rotary inlet ports, working in conjunction with the hollow and ported central journal of the crankshaft. By this arrangement the inlet period of induction from carburettor to crankcase could be timed to the best advantage, thereby obtaining improved volumetric efficiency.

The efficient Schnuerle loop-scavenge and flat-top piston system was in increasing favour on the Continent, and was incorporated in several designs such as the Ilo, Jawa, Püch, D.K.W. among others. The small-capacity two-stroke motorcycle tended to become more elaborate and to be used for a greater variety of purposes, thus indicating that the average size and weight of machines were becoming less as specific power outputs and reliability increased. An example of this tendency was the 199 c.c. N.S.U. 'Lux' design, which had a light spine type of frame and an enclosed oil-bath case for the rear chain. This latter item was an indication that the oil-bath chain case was coming back into use as an additional assistance to the chains, which were now stressed to a higher extent because of the continually increasing speeds and specific powers of engines.

Two-stroke design was very vigorous on the Continent after 1950 and resulted in the production of such designs as the 98 c.c. Guzzi 'Zigolo' which, with a rotary inlet valve, developed the remarkable output of 6·8 h.p. at 8400 r.p.m.; the 200 c.c. Iso twin-piston engine which developed 10·5 h.p. at 4550 r.p.m.; and the German 250 c.c. Triumph T.W.N. four-cylinder, twin-piston engine which developed 14 h.p. at 4800 r.p.m. The 350 c.c. Ehrlich twin-piston engine (Plate 13) was similar to the Püch design. A still more complex multi-cylinder two-stroke engine was the air-cooled three cylinder radial design produced by the German D.K.W. firm, initially for racing purposes. Such advances made possible the establishment of the ultra-lightweight and lightweight types of motorcycle for all-round duties which had formerly been performed by machines of double and treble the capacities. Of 116 different models of motorcycles produced in 1953 by twenty-four German manufacturers, 99 were equipped with two-stroke engines.

MOTOR SCOOTERS AND MONOCARS

The conceptions of the simple motor scooter and the more elaborate monocar, which were originated in the 1920's but were not then accepted as practical, became established after 1945 when it was demonstrated on the Continent that these specialised forms of motorcycles could now be used reliably for solo or passenger-carrying runabout and touring purposes. Fitted with engines of from 98 to 200 c.c. capacities, and weighing from 110 lb. to 180 lb., these handy vehicles provided convenient and stable transport for one or two people at speeds up to 40 m.p.h. and at consumptions of some 120 miles to the gallon of fuel. This general handiness, combined with adequate performance and economy in operation, created a popular appeal under the existing circumstances of financial stringency and high costs. The success of the motor scooter, particularly as a baggage-carrying two-seater, and even with a separate baggage trailer or with a sidecar, eventually caused its elaboration to be considered in the form of a monocar, which would include full weather protection as well as other amenities.

One of the earliest and simplest designs of these new motor scooters was the Corgi, which mounted a 98 c.c. Villiers two-stroke engine horizontally in a simple tubular frame supported on two small wheels of $12\frac{1}{2}$ in. diameter. It was introduced as a portable transport for Army parachutes. The chain drive was taken to the rear wheel through a 2-speed gearbox and hand-operated clutch. Handlebar steering and a bicycle-type saddle were used. So useful did this simple form of personal motor transport become, that a great number of these machines were sold in the first years of their manufacture.

The two chief Continental designs, which were quickly adopted and widely used after 1945, were the Vespa and the Lambretta, both of Italian origin. The Vespa, made under licence in England by the Douglas Company, fitted a 125 c.c. single-cylinder two-stroke engine which developed 4 h.p. at 4500 r.p.m. It was equipped with a 3-speed gearbox, clutch and kick-starter, front and rear springing, electric lighting and adequate weather protection.

The Lambretta scooter was of similar but somewhat more elaborate design, having a 123 c.c. engine (Plate 14) which developed 5 h.p. at 4600 r.p.m. and a shaft drive to the rear wheel. The seating arrangements on these more elaborate designs made use of the enclosure of the rear portion of the machine, and used the top of this covering to attach one or two padded seats of greater comfort than the simple bicycle-saddle type. Provision was also made for the mounting of a spare wheel. In addition, elaborate leg and wind shields, as well as mudguards, were incorporated, thus affording a high degree of weather protection, which added substantially to its suitability as an all-purpose machine for use either as an occasional runabout or a vehicle for more serious travel. Another advanced Continental design was the German 150 c.c. Zündapp 'Bella', whose engine developed 7·3 h.p. at 4700 r.p.m. and which was fitted with a 4-speed gearbox. With a top speed of over 50 m.p.h. and seating for two, this design illustrated the progress that had been made in motor scooters by 1953.

The British-designed and built Oscar machine was still more elaborate and powerful, being fitted with engines of either 125 or 200 c.c. capacities, the latter of which developed 8·4 h.p. at 4000 r.p.m. It had seating for two people and was capable of speeds up to 45 m.p.h.

AUTOCYCLES

The autocycle had already become established as an economical form of personal motor transport before 1939, and in the demand for economy after 1945, it continued to be accepted as a practical ultra-light motorcycle. Its general form remained fairly constant, having the high open form of diamond frame equipped with a 98 c.c. two-stroke engine unit, either with or without a 2-speed gear, a hand clutch and also pedalling gear. Spring forks and large-section tyres were now standardised in the interests of greater riding comfort, and the engine unit was generally enclosed in some form of weather-proof cowling.

An interesting development of this simple form of motorcycle was the Scott 'Cyc-Auto' of 1953 which, with a 98 c.c. two-stroke engine mounted in a simple open tubular frame, also included the unusual feature of rear springing, a unit construction 2-speed gearbox, and shaft drive to the rear wheel. The autocycle of this period tended to merge with the ultra-light motorcycle proper with a small yet relatively powerful engine, light frame and small-diameter, large-sectioned tyres.

CYCLEMOTOR UNITS

After 1945, economic pressure revived the need for ultra-cheap mechanical transport and an increasing number of new cyclemotor designs appeared on the Continent and in Great Britain. The old problem of where to put the motor unit on the cycle still existed, and the new types were nearly as varied as their predecessors had been at the beginning of the century. Two of the most convenient forms were the self-contained motorised wheel, which conveniently replaced the normal rear wheel of the cycle, and the 'clip-on' type of unit that was attached under the pedalling bracket and drove the rear-wheel tyre either by means of a friction wheel or through a chain.

These new designs were much smaller and operated at much higher crankshaft speeds and specific power outputs than their counterparts of fifty years before. Reliable units of from 25 to 50 c.c. capacities, developing ¼ to 1 h.p. at crankshaft speeds up to 5000 r.p.m., and weighing from 15 to 20 lb., were available in 1950. Fuel consumptions of up to 200 m.p.g. at speeds of up to 30 m.p.h. were obtained with these modern units.

A good example of the modern self-contained motorised wheel was the 'Cyclemaster' (Inv. 1952–199). This unit had a two-stroke engine of 30 c.c. capacity, which drove the wheel through chains and a hand-controlled clutch. It weighed 19 lb. and was able to cover 250 miles on a gallon of petrol. Another well known make of this type was the 35 c.c. B.S.A. 'Winged-Wheel' unit.

The Trojan 'Minimotor' unit (Inv. 1952–290) was designed to be fixed over the rear wheel of the bicycle and drive the tyre by a grooved friction wheel. It had a cylinder capacity of 49·9 c.c., and weighed 24½ lb. Its comparatively powerful engine permitted road speeds of up to 30 m.p.h. and it was able to cover 150 miles to a gallon of petrol. Various other types, such as the Berini and the Power Pak units, were of generally similar type and intended to drive either the front or rear tyres by a friction wheel. The 50 c.c. engine of the Solex cyclemotor was mounted on the steering head and drove the front wheel by means of a friction roller in contact with the tyre tread.

When the cyclemotor unit was clipped to the cycle frame under the pedalling bracket, a compact installation and a low centre of gravity resulted which improved stability. The 'Mosquito' unit, of Italian manufacture, was fixed in this manner and drove the rear tyre through a friction wheel; its 38·5 c.c. two-stroke engine propelled the cycle for 200 miles on a gallon of petrol, and the complete unit weighed only 15 lb. The Italian 'Cucciolo' unit (Inv. 1952–198) was also attached under the pedalling bracket, but had a more complex and elaborate specification. Its high-efficiency overhead-valve four-stroke engine was of 48 c.c. capacity, and it had a 2-speed preselector gear, multi-plate clutch and a positive drive to the rear wheel through the normal pedalling chain. In spite of its complications, the 'Cucciolo' unit weighed only 17½ lb.; it developed 1½ h.p. at 5300 r.p.m. and consequently had a high road performance.

One of the most interesting developments in cyclemotor designs was the German 18 c.c. Lohmann two-stroke compression-ignition unit. It was designed to clip on the cycle frame under the pedalling bracket and drive the rear tyre by means of a friction wheel; it weighed only 11 lb., and developed ¾ h.p. at 6000 r.p.m. It was able to cover 350 miles on a gallon of fuel at speeds up to 20 m.p.h.

Various other designs embodying rotary engines, multi-cylinder engines, belt drive and other unorthodox features were also produced, but they were in less favour than the types described.

ACCESSORIES

Detail improvements in the design of accessories which were introduced after 1945, made refinements rather than major changes in general design. Frames remained for the most part basically of the elaborated diamond form, but they were invested with greater strength, rigidity and operational stability particularly for high speeds, by the use of duplex tubes in cradle form. An example of this process was the arrangement known as the Norton 'Featherbed' frame.

Frames of pressed steel, in which weight was saved and weatherproofing improved, were made by forming the rear portion as a stressed mudguard and adding valances capable of taking the main structural stresses. A central backbone or spine type of frames, similar in principle to the bicycle 'crossframe' of the 1880's, was also adopted by a few makers of lightweight machines as a means of simplifying manufacture and saving weight. The use of spring frames increased, not only for the heavier and more powerful machines, but also for many lightweights with engine capacities as low as 98 c.c. The two main forms used were the plunger coil spring and the swinging rear forks; by 1953 the latter system was the more generally adopted. Tyre sections were made still larger to improve riding comfort.

The use of the tele-hydraulic type of spring-fork increased rapidly after 1945 because of the improved riding comfort and steering qualities that it provided. This form was challenged in 1952 by the Earles swinging-link type of spring fork, which appeared to owe much to the excellent Truffault spring fork of 1904. The Earles design was claimed to provide better steering qualities at high speeds than any other form, although it was more complicated and heavier than the tele-hydraulic type.

A significant change of these post-war years was a tendency to adopt either an oil-bath chain case for the rear chain or an all-enclosed shaft drive in place of the exposed chain drive which, since the 1920's, had been in general use. The still higher power outputs of engines were beginning to reveal inadequacies in the chain, particularly under high-powered single-cylinder racing conditions. Serious consideration was once more given to the adoption of shaft drive as was shown by its use in such advanced designs as the Sunbeam S.7, the L.E. Velocette and the B.M.W. By 1953 some ultra-light machines were also adopting shaft drive.

Electric-lighting systems were improved by such innovations as the adoption of the alternating-current form of generator and the incorporation of the headlight within the handlebar and steering-head assembly. Sidecars improved in general construction, and in such details as weatherproofing, mudguarding, upholstery, suspension and luggage-carrying capacity.

SOCIAL AND SPORTING DEVELOPMENTS

In spite of the fact that the Brooklands race track was dismantled in 1939, thus depriving England of its sole facility for high-speed motor car and motorcycle events, the popularity of track racing took firm hold of the younger generation after the war. Alternative, improvised sites for such events were conveniently found in the perimeter tracks of various extensive airfields which had been constructed during the war, such as Silverstone, Goodwood, Snetterton and some others.

These, together with the classic Tourist Trophy Races held annually in the Isle of Man, and the many important Continental road-racing events, received wide support. The establishment of such amateur events as the

Manx Grand Prix and the Clubman's Tourist Trophy Races, both of which were run over the mountain course in the Isle of Man, catered still more directly for amateur enthusiasts.

With the aid of ever-improving designs, world records continued to be improved. The lap-record of the Senior Tourist Trophy Race was raised in 1955 to 99·97 m.p.h., and the world speed record for a 1000 c.c. motorcycle stood at 185·15 m.p.h. In addition, remarkable speeds with some of the smallest engines were established, such as the speed of $124\frac{1}{2}$ m.p.h. attained with a 123 c.c. two-stroke engined machine. The successful circling of the globe was achieved with a 49 c.c. cyclemotor-equipped bicycle in 1953. In addition, such events as the English and Scottish Six Days' Trials, International Six Days' Trials, London-to-Land's-End runs, pioneer runs, scrambles (moto-cross) and other events also acquired great popularity both here and on the Continent. Similar events, run mostly with highly-efficient British machines, also began to create interest in some parts of the United States.

Miniature Motorcycles: from 1956

From the aspect of the high-speed internal-combustion engine, the decade of the 1960's was notable for the considerable progress achieved with its detail technical design and substantial increase in specific outputs with single and multi-cylinder engines of both the two-stroke and four-stroke varieties. This substantial technical advance, made possible by increased investment in sophisticated research methods, new materials of improved properties, closer manufacturing tolerances standardized by rigorous inspection routines, ever improving manufacturing techniques, and substantiated by intensive and prolonged testing under both racing and extended touring and trials conditions of operation, was applied to all sizes and categories of motorcycle power units. This progress, continuing from the beginnings in the 1950's of significantly increasing specific outputs already noted in Chapter 7, began to render obsolescent the classic medium and heavy forms of motorcycle which had been generally accepted during the previous four decades, and cause them to be partially replaced by a variety of newer, smaller, lighter and more economical designs of equal or even greater comparative performance.

In respect of the small motorcycle of from 50 c.c. to 250 c.c. capacities in particular, this technical advance sponsored progressive new forms of motorcycle owing little to the classic diamond frame configuration which, since the 1890's, had been standard practice for both bicycles and motorcycles. These new forms, foreshadowed in the tentative designs of the 1950's, were the cyclemotor or moped, the motor scooter and the advanced ultra-lightweight motorcycle proper. These three new distinctive forms were now subjected to vigorous technical development and commercial exploitation and, under the impetus of wide popular acceptance, were put into intensive and commercially successful mass production in Germany and Italy and later in England and Japan not only by numerous of the most important existing motorcycle manufacturers, but also by highly progressive newcomers, particularly in the case of Japan. The sudden rise and quick success of Japanese motorcycle engineers were notable phenomena of the 1960's.

The fresh design characteristics which distinguished these new forms, apart from the substantially higher specific outputs of their power units, were the increased simplicity and lightness of the open single spine type of frame used generally with variations for the cyclemotors and miniature motorcycles, and the open spine or platform frame of the motor scooter. These new design trends, developed from their beginnings in the 1950's, resulted generally in more convenient and easier riding characteristics – eliminating for good, for instance, the former distinction between a 'lady's' and a 'gentleman's' mount, and providing simpler and cheaper mass-production methods, such as, for instance, the quickly and cheaply produced pressed steel and welded-spine frame.

To this originality in frame design were added the remarkably high specific outputs of reliable power units already noted, and considerably improved detail design including such features as powerful drum and even disc brakes which, in some of the largest sizes, were hydraulically operated, and robust brake controls and fitments which were a useful improvement on the somewhat flimsy fitments previously generally employed; efficient adjustable front and rear sprung suspensions which gave comfortable riding in spite of the smallness of the mounts; three, four and five-speed gear boxes; simple, automatic transmissions; smooth, durable manually-operated clutches and, in some cases, automatic clutches even with the smallest mounts; full electric equipment comprising lighting, direction-indicators, brake-stop lights, supply for ignition requirements and, in a few cases, electric engine starting; adequate weather protection, especially on motor scooters whose inherent quasi-monocar form contributed to this protection; and improved road holding and general safety of control.

Still further remarkable progress was achieved in the case of special racing designs of small normally aspirated engines of up to 250 c.c. capacity having from one to six cylinders and capable of producing up to 200 h.p. per litre capacity at crankshaft speeds of up to 19,000 r.p.m. at continuous full-throttle operation with sustained reliability, as was consistently demonstrated by many successful performances in arduous international racing events of all kinds. Such spectacular technical advance may be illustrated by the 50 c.c. capacity two-stroke engine performance which won the 1964 50 c.c. Tourist Trophy at an average speed in excess of 92 m.p.h. Such experimental progress was potentially available to future production designs.

Although there were a number of representative British designs on the market in the 1960's which conformed to these new concepts of what may be collectively called miniature motorcycles, the origin of them was largely foreign. The cyclemotor owed much to German ingenuity which supplied the need after 1945 for very economical and convenient personal transport; the modern motor scooter, although having such tenuous antecedents as the Ellehamer prototype of 1905 and the ABC Scootamota and Jebb production designs of 1919, was largely Italian in concept; and the miniature motorcycle, although existing in low-powered forms since the 1920's, was revolutionized by Italian and later still further by Japanese engineers. With their economy of operation, high relative performance and unfailing reliability, all three types had an immediate and world-wide consumer appeal, with the result that with modern mass-production methods they were soon flooding world markets.

It is to be noted that this new extensive development and production of miniature motorcycles during this period made obsolete the simple engine units for attachment to pedal bicycles which had had some popularity during the previous two decades.

CYCLEMOTORS

The cyclemotor or moped, more properly named the motor-assisted bicycle since its low power needed occasional pedal assistance, stems back to the earliest days of the motorcycle when most machines were of generally similar specification. Development of this type proceeded through such pedal-assisted designs as the Autowheel, J.E.S. and Dayton motor-attachment units of 1914, and the small two-stroke engined motorcycles of the 1930's such as those which fitted the 98 c.c. capacity Villiers engine. The need for cheap personal transport after 1945, particularly in impoverished Europe and Japan, revived the concept of the cyclemotor. The considerably improved design and manufacturing techniques which were now available aided the production of an improved form of cyclemotor having relative high performance combined with dependable reliability, and an economy of operation resulting from a low road tax and a petrol consumption as low as 200 m.p.g. The majority of cyclemotors were equipped with a single-cylinder air-cooled two-stroke engine of about 50 c.c. capacity which developed between $1\frac{1}{2}$ to $2\frac{1}{2}$ h.p. at crankshaft speeds of around 5000 r.p.m. Japanese designs of this category developed nearly twice this output at crankshaft speeds of up to 10,000 r.p.m. The standard of performance of some of these higher powered designs, particularly in the case of those from Japan, made pedal gear unnecessary.

The simplest form of cyclemotor was represented by the long established Solex design which was essentially an open-frame pedal bicycle of pressed-steel construction, the front forks of which carried a 49 c.c. (39·5 mm. by 40 mm.) air-cooled two-stroke engine unit positioned over the front wheel tyre tread which it drove by a friction roller. The clutch action was automatic. The 50 c.c. Clark Scamp cyclemotor had a simple open tubular frame and a motorized rear wheel.

One of the earliest of this advanced type of cyclemotor was the N.S.U. Quickly design, which was internationally popular over a long period and was produced in various models in large numbers. The 49 c.c. capacity single-cylinder air-cooled two-stroke engine developed 1·4 to 2 h.p. at 5,500 r.p.m., depending upon the degree of tune intended for various models. A 2 or 3-speed gearbox integral with the engine crankcase transmitted the drive to the rear wheel through an open chain. The 50 c.c. capacity Raleigh Runabout and Ultramatic models were similar to the French Mobylette; this design had the progressive features of an unlined light-alloy cylinder with a hard chromium-plated bore which provided better heat transfer and lighter construction, a flexible vee-belt primary drive which gave a smooth and quiet drive, an automatic clutch and, in the case of the Supermatic model, an infinitely variable automatic gear change. The Raleigh Supermatic model was equipped with a split spring-loaded driving pulley which opened as the load on the belt increased and permitted the belt to sink to a lower effective pulley diameter and so reduce the gear ratio, and to rise to a higher effective diameter to raise the gear ratio as the belt tension decreased with the lessened load. The engine was pivot mounted at the cylinder head to maintain correct belt tension during gear ratio variation. Tuned versions of this design developed 2·6 h.p. at 5,600 r.p.m. and were capable of a top speed of 45 m.p.h. and a fuel consumption of 120 m.p.g. An advanced cyclemotor design was the Raleigh Wisp (Plate 15) which used the same basic engine unit. The frame was of open tubular spine construction supported on small-diameter wheels fitted with large cushion tyres in lieu of a spring suspension system. This simple type of frame was similar to the contemporary Raleigh pedal bicycle which followed the original Moulton design of pedal bicycle, and which was itself foreshadowed by the cross-frame safety bicycle of 1886. The fully-automatic transmission, compact size and light-weight made this original cyclemotor a handy and economical runabout.

Such was the commercial viability of this new and efficient form of cyclemotor that nearly all of the most prominent motorcycle manufacturers, as well as certain smaller new ones, produced their own designs which conformed generally to the simple specification already described. Among the 50 c.c. category may be mentioned the 47 c.c. Binetta; 50 c.c. D.K.W.; 48 c.c. Durkopp Dianette; 50 c.c. Garelli; 50 c.c. Gilera; 49 c.c. Hercules; 50 c.c. Jawa; 49 c.c. Kreidler Junior; 39 c.c. Lambrettino; 50 c.c. MV; 50 c.c. Moto Guzzi; 49 c.c. Norman Nippy; 49 c.c. Peugeot; 49 c.c. Püch; and 49 c.c. Zündapp. In the 100 c.c. category were such designs as the 94 c.c. Capri; 75 c.c. F.N.; 98 c.c. James; 98 c.c. New Hudson; and 70 c.c. Zundapp. Specialized cyclemotor engine units were also produced by such large proprietary firms as Villiers and Sachs.

Japanese examples of cyclemotor designs were remarkable for their original layout, full operational and weather shield equipment and outstandingly high specific outputs. The 50 c.c. (41 mm. by 38 mm.) capacity Suzuki M30 design (Plate 15) had a simple two-stroke engine which developed 4 h.p. at 6,800 r.p.m. The drive was taken to the rear wheel through a 3-speed integral gearbox which incorporated an automatic clutch having multiple plates running in oil, and a secondary driving chain enclosed in a dust-proof case. Powerful over-size drum brakes were fitted, and the suspension comprised sprung front forks with oil damping, and sprung swinging-arm rear suspension with oil damping. It will be noted that this specification was more comprehensive than that of the simpler European designs already mentioned. The 50 c.c. Yamaha Scooterette design was similarly progressive in its technical features and road performance. A more powerful version of this Yamaha cyclemotor was fitted with a 73 c.c. (47 mm. by 42 mm.) single-cylinder two-stroke engine, having a rotary inlet valve and automatic lubrication, which developed 6·2 h.p. at 7000 r.p.m.

The Honda concern specialized on the four-stroke engine for powering their wide variety of advanced motorcycle models. In the cyclemotor range, the Honda 50 design (Plate 15) had a general specification similar to the Suzuki M30, but was fitted with a 49 c.c. (40 mm. by 39 mm.) capacity four-stroke air-cooled engine with overhead push-rod operated valves (Fig. 12), which developed 4·5 h.p. at 9,500 r.p.m. This exceptionally high crankshaft speed and resulting high specific output were characteristic of the new Japanese designs which, after 1960, quickly captured world-wide markets in competition with European and British manufacturers. Automatic clutch, 3-speed gearbox, front and rear spring suspension, coil ignition and electric lighting supplied from an engine-driven generator, and efficient weather guards were features of this design. A more unusual Honda design was the Monkey Bike Model which used the same 49 c.c. four-stroke engine unit and transmission. This model was a miniature motorcycle on the general lines of the early Corgi and was intended as a compact auxiliary personal transport which could conveniently be carried in the boot of a car or aboard a yacht. It weighed only 119 lb. This same 49 c.c. capacity over-head valve engine was also used in the self-contained motorized rear wheel arrangement of the Little

Honda design, which had an open unsprung welded pressed-steel spine frame, automatic clutch and pedalling gear. Luggage racks front and rear provided an additional 60 lb. of luggage weight capacity; with the rider making a total of some 200 lb., this little machine was capable of covering up to 230 miles on a gallon of fuel at an average speed of 20 m.p.h.

The Honda CM90 model was a more powerful version of the Honda 50, having a 90 c.c. (50 mm. by 45·6 mm.) single-cylinder, overhead-camshaft engine which developed 7·5 h.p. at 9500 r.p.m. with a compression ratio of 8 to 1.

MOTOR SCOOTERS

The antecedents of the modern motor scooter included such tentative designs as the 1916 Autoped, the 1919 A.B.C. Scootamota and Kingsbury, the 1921 Autoglider, and the 1928 Harper. Although such primitive prototypes evoked the basic concept of the motor scooter, they did not succeed in establishing it in commercial competition with the generally accepted classic form of motorcycle. It was not until 1945 that the motor scooter appeared in a technically advanced form which founded a new concept of the motorcycle and created an entirely new market among the young who wanted cheap reliable personal transport, without having the technical interest in their mounts characteristic of the adherents of the classic form of motorcycle. The prototype of the modern motor scooter was the Vespa ('Wasp') designed by C. d'Ascanio of the Italian aircraft firm of Piaggio and Co. By April, 1946, this fundamentally new concept had gone into quantity production, and by 1960 over two million had been sold and the market for it had been established among a new and rapidly expanding clientele.

The Vespa design owed much to aircraft engineering from which it had proceeded. A stressed-skin spot-welded spine frame was used; the wheels were mounted on stub axles, thus making the wheels easily detachable without having to disturb the final drive and so simplifying the problem of punctures. A simple 98 c.c. capacity two-stroke engine was adopted and mounted midframe and totally enclosed in detachable side panels, and a sheet-steel platform framework attached to the central spine and extended upwards almost to the handlebars provided both support for the feet and weather protection. The drive to the rear wheel was by a simple chain from the engine shaft to a 3-speed and clutch unit mounted adjacent to the rear wheel axle. This Vespa direct drive system evolved with the prototype proved to be the best and most generally adopted arrangement for motor scooters, in spite of various other forms of drive – belt, shaft, etc. – which have been tried in other designs. The rapid expansion of production of Vespas at the Piaggio works was complemented by their manufacture under licence in Germany, England and France. Newer models, but still based on the prototype design, were produced with larger engines up to 150 c.c. capacity.

In 1948, Innocenti of Milan produced a motor scooter which had a like commercial success. This differed in detail from the Vespa design in such items as front forks, a duplex tubular cradle frame, lower front weather guard, foot-operated gear change, and shaft drive; moreover the prototype was provided with a pillion seat. The engine and drive was not enclosed in side panels. The Lambretta was manufactured in Germany and France and later imported in large numbers to England. The success of this second Italian design thus complemented and confirmed the original success of the Vespa.

This combined Italian progress provided the impetus which set in motion a wide emulation by other manufacturers, and by 1960 a considerable variety of motor scooter designs appeared having, in the majority of cases, two-stroke engines of from 50 c.c. to 200 c.c. capacities, and such embellishments as 4-speed gearboxes and, in some cases, automatic clutches and even automatic transmissions, electric engine starting, and ingenious and commodious accommodation for luggage and spare wheel as well as a pillion passenger. These mounts were capable of long distance touring with full load of driver, passenger and luggage of a total weight of up to 300 lbs., with fuel consumptions of between 90 and 150 m.p.g. Moreover, adequate leg and handlebar shields and tall transparent windscreens together provided a high degree of weather protection which contributed still further to the practicability of this new form of miniature motorcycle. Some of the largest engines, up to 300 c.c. capacity, were of the more elaborate and powerful two-cylinder four-stroke design. Such engines had outputs of up to 12 h.p. at crankshaft speeds of around 5000 r.p.m. Maximum speeds ranged from 30 to 67 m.p.h., and fuel consumption from more than 200 to 90 m.p.g. respectively for these two extremes of design.

The unorthodox layout of the modern motor scooter necessitated specialized applications of direct, chain, belt and shaft transmissions to be used. Direct drive was adopted in Vespa designs from the beginning. The crankshaft of the Vespa engine, which was placed as close as possible to the rear wheel, was at right angles to the centre line of the frame and the drive to the rear wheel was through a clutch and cush drive and straight-cut spur gears, which were totally enclosed and ran in oil. Another form of direct drive was used in the N.S.U. Prima design in which the engine was placed across the frame so that the crankshaft ran parallel to the axis of the frame. The drive was taken through a clutch and gearbox, as in a car layout, and the short extension shaft of the gearbox drove the rear wheel hub through a pinion and bevel gear. With chain drive, the crankshaft drove through sprockets and chain to a clutch and gear assembly incorporated adjacent to the rear wheel spindle. This form of transmission was adopted in the Lambretta models (Plate 14), after the shaft drive of the early models has been abandoned. Belt drive, with its essential simplicity and smooth transmission, which was used with such low-power scooters as the 74 c.c. DKW Hobby and, the Concorde and the designs, was arranged to constitute a simple yet effective automatic variable transmission. This Uher system employed spring-loaded split driving and driven pulleys; when the load on the driving pulley increased, the pulley flanges were pushed further apart and the belt sank to a smaller effective diameter, thus lowering the gear ratio. At the same time, the load on the driven pulley diminished because of the slackening of the belt, and the spring-loaded flanges tended to come together, thus increasing the effective diameter of the driven pulley. These complementary actions resulted in a lowering of the overall gear ratio. A lessening of the road load had the converse effect and the

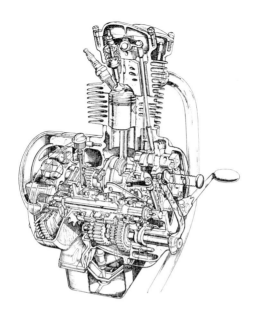

Fig. 14   50 c.c. Ariel 'Pixie' engine: 1963

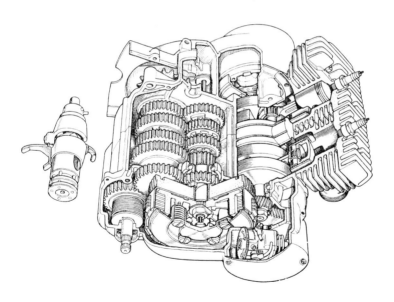

Fig. 15   97 c.c. Yamaha twin-cylinder engine: 1966

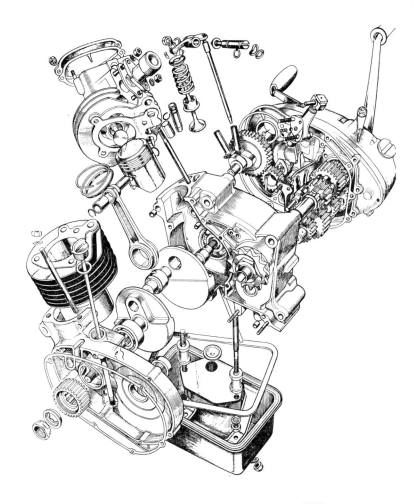

Fig. 16   75 c.c. B.S.A. 'Beagle' engine: 1964

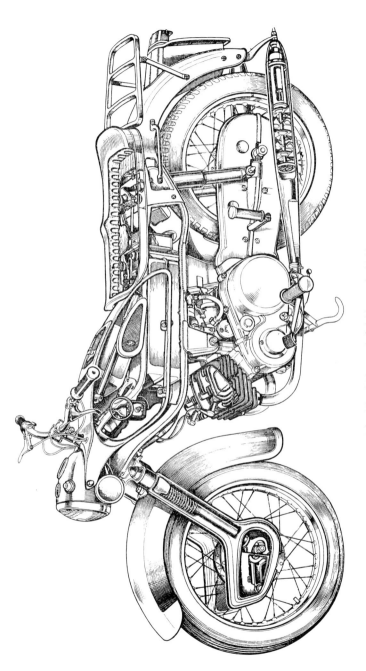

Fig. 17 250 c.c. Ariel 'Arrow': 1959

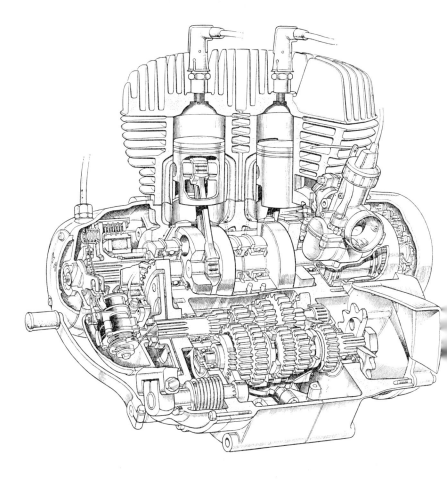

Fig. 18   246 c.c. Yamaha twin-cylinder engine: 1965

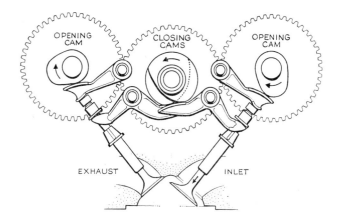

Fig. 19 Ducati desmodromic valve gear: 1960

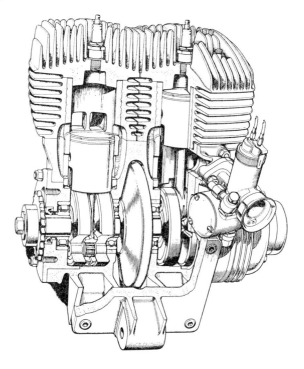

Fig. 20 344 c.c. Scott twin-cylinder two-stroke engine

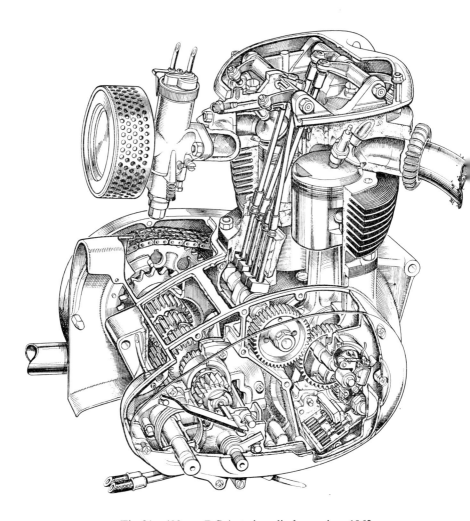

Fig. 21   499 c.c. B.S.A. twin-cylinder engine: 1962

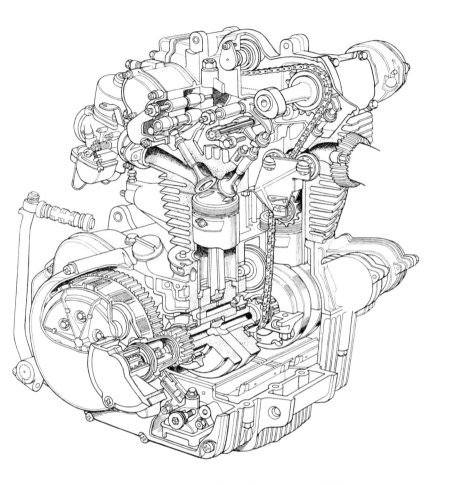

Fig. 22 444 c.c. Honda twin-cylinder engine: 1966

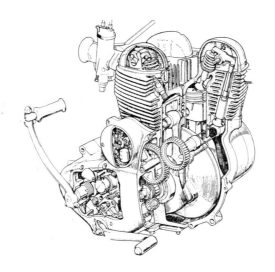

Fig. 23   250 c.c. Norton twin-cylinder engine: 1958

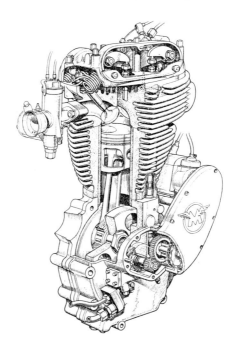

Fig. 24   500 c.c. Matchless engine: 1948

overall gear ratio was raised accordingly. The pulleys were thus constantly adjusting to varying load conditions to provide optimum operational gear ratio settings.

Some other forms of transmission have been applied to motor scooters such as shaft drive and even a fully-automatic drive through a hydraulic torque converter. Shaft drive, used initially with Lambretta and N.S.U. machines, was little used because of higher manufacturing costs; one of the few designs incorporating shaft drive was the 248 c.c. Velocette Viceroy which had a twin-cylinder horizontally-opposed two-stroke air-cooled engine. The hydraulic automatic drive was used in production in Japan, and in prototype form by Ducati in Italy and N.S.U. in Germany. The high cost and power losses inherent in this system make it too costly for application to the motor scooter.

There were, by the end of the 1960's, a great number and variety of motor scooter designs with engine capacities of from 50 c.c. to 300 c.c. capacities, which were mostly of the two-stroke single-cylinder kind with a few of the twin-cylinder four-stroke kind in the largest sizes. The following paragraphs summarize the main categories in terms of engine capacity and mention some of the more important and distinctive design features in each category, together with some of their more individual details to present a broad picture of technical development as it was applied to actual production.

The smallest category of motor scooters had simple two-stroke single-cylinder engines of up to 50 c.c. capacity, which developed about 3 h.p. at 5000 r.p.m. Although naturally limited in performance and carrying capacity, yet they formed a useful link between the cyclemotor and the larger motor scooter, having the lightness of the former and the advantages of the advanced design features, such as low centre of gravity and weather protection, of the latter. Although such designs were of limited performance, yet most of them were equipped with a second saddle for passenger accommodation. This small category included the 47 c.c. Binz; the 50 c.c. Jawa; the 49 c.c. Kreidler; the 50 c.c. Mobylette; the 50 c.c. Motobi and the 47 c.c. Victoria. All these designs were fitted with simple two-stroke engines. The 49 c.c. Dunkley Popular Model had a more elaborate overhead-valve four-stroke engine, later increased to 65 c.c. capacity, which gave it a better performance in comparison with the two-stroke engined models mentioned. The later 50 c.c. Lambretta Luna model was an advanced design, styled by Bertone, whose engine developed 2·5 h.p. The 75 c.c. Cometa model of similar design was fitted with an engine which developed 5·5 h.p. and was lubricated by an oil-injection system.

The next larger category of up to 100 c.c. capacity included the 98 c.c. Vespa prototype of 1945 already mentioned; this capacity was soon increased to 123 c.c. for production machines to provide a more useful performance. Twenty years of intensive commercial production since then, however, have materially improved the performance and quality of such low engine capacity machines and a variety of later models have appeared which were successfully used for both town work and long distance touring. Among the earlier examples were the 74 c.c. Manurhin, formerly known as the D.K.W. Hobby, which had the distinction of an automatic clutch and a variable transmission which worked with expanding and contracting spring-loaded pulleys and primary belt drive; the rider controlled the speed with the twistgrip throttle and brakes only. The 70 c.c. B.S.A. Dandy was fitted with a two-speed preselector gear; the 98 c.c. Adler Junior had a 3-speed gearbox and rear enclosed chain drive and an electric starter. The 70 c.c. Capri was fitted with a Garelli 45 mm. by 44 mm. two-stroke engine which developed the high output of 3·3 h.p. at 6500 r.p.m., and this substantial output was further assisted by a 3-speed gearbox. The 60 c.c. (42 mm. by 43 mm.) Püch Cheetah design had a 3-speed gearbox and a two-plate clutch running in oil; the primary drive was by a chain enclosed in a pressed-steel case. The 78 c.c. (48 mm. by 43 mm.) Raleigh Roma was derived from a Bianchi design; it was fitted with a Marelli 6-volt flywheel generator with lighting coils and remote high-tension coil; a 3-speed gearbox, multi-plate clutch running in oil, helical primary gear and rear chain enclosed in a plastic case transmitted the drive to the rear wheel. The Italian 75 c.c. Motorbi Scooterette design had a compact overhead-valve four-stroke engine, 3-speed gearbox and wet clutch unit mounted just forward of the rear wheel, which was driven through a chain running in an oil bath case. Later more advanced models in this class included the 98 c.c. Lambretta Cento which developed 4·7 h.p. at 5500 r.p.m., and had a top speed of 47 m.p.h., a fuel consumption of 150 m.p.g., a 3-speed gearbox, a multi-plate clutch running in oil and an enclosed rear chain case; the 100 c.c. Triumph T10 design had an automatic clutch and transmission; and the 88 c.c. (47 mm. by 51 mm.) Vespa Super Sprint 90 (Plate 16), equipped with a 4-speed gearbox in unit with the engine crankcase, a lightweight stressed-steel monocoque chassis, and a spare wheel carried conveniently within the loop of the frame. The petrol-oil ratio for lubrication required for this Vespa engine was only 50 to 1.

The next higher class of motor scooters with engine capacities up to 150 c.c. was represented initially by the 124 c.c. (54 mm. by 54 mm.) Vespa prototype production model of 1948. The engine-speed gear-transmission unit was compactly mounted on the right hand side of the rear wheel; the 3-speed gear and multi-plate clutch assembly took the drive from the engine shaft to the rear wheel hub through a train of straight-cut gears. This original Vespa transmission layout has been retained for all subsequent models, thus demonstrating the soundness of the original design. The later Vespa 150 model having a capacity of 145 c.c. (57 mm. by 57 mm.) was capable of a top speed of nearly 60 m.p.h. and had a cruising fuel consumption of 100 m.p.g. The efficient two-stroke engine had a deflector piston and a rotary inlet valve, the latter improving volumetric efficiency and consequently power output, a notable feature of this model.

The early 123 c.c. (52 mm. by 58 mm.) Lambretta production design employed a 3-speed gearbox in unit with the engine, a multi-plate clutch running in oil, and a final drive to the rear wheel hub by bevel gears and shaft enclosed in a pivoted rear stay; it had a top speed of 42 m.p.h., a fuel consumption of 113 m.p.g., and a total weight of 243 lb. The 148 c.c. (57 mm. by 58mm.) LAD model had the refinement of a Bendix-type electric starter. The shaft drive transmission was abandoned and replaced for subsequent models beginning with the Li and TV models by duplex chain drive from the engine shaft to a clutch and 4-speed gear unit located adjacent to and driving the rear wheel hub. The Li 125 was the first of this new design and was the forerunner of the more advanced 148 c.c. (57 mm. by 58 mm.) Li 150 model, whose increased power gave it a speed of 52 m.p.h.

A considerable number of 125 c.c. designs on the general fashion set by the contemporary Vespa and Lambretta equivalents were produced by various continental manufacturers including: the 123 c.c. Motobecane Mobyscooter; the 125 c.c. Piatti; the 123 c.c. TWN; the 124 c.c. Terrot with 3-speed preselector gearbox; the 124 c.c. Rumi, which had the unusual feature for this low power machine of a twin-cylinder two-stroke engine driving through a 4-speed gearbox; the 123 c.c. Adler which was equipped with an electric starter; and the 125 c.c. Heinkel which had an overhead valve four-stroke engine fitted with an electric starter. The 150 c.c. designs included the Püch Alpine of 147 c.c. (57 mm. by 57 mm.) which developed 6 h.p. at 5500 r.p.m.; the N.S.U. Prima of 147 c.c. (57 mm. by 58 mm.) which developed 6·2 h.p. at 5000 r.p.m. and was fitted with a flywheel starter generator; the 125 c.c. Peugeot; and the 148 c.c. Zündapp Bella. One of the most advanced designs in the 150 c.c. class of motor scooters was the Lambretta SX 150 (Plate 16) having a powerful 148 c.c. (57 mm. by 58 mm.) engine which developed 9·4 h.p. at 5600 r.p.m., and gave this machine a maximum speed of 58 m.p.h. and a fuel consumption of 110 m.p.g.

The still larger designs of motor scooter with engine capacities up to 200 c.c., while they were more powerful and of higher performance and weight carrying capacity, still followed the same basic design form as laid down in 1945. This larger class was well provided with a variety of both Continental and British designs which developed outputs of up to 10 h.p. and had maximum speeds of 60 m.p.h.; moreover, their higher performance introduced such features as disc brakes. The normal type with single-cylinder two-stroke engines included the earlier 160 c.c. and 180 c.c. Vespa and 175 c.c. and 180 c.c. Lambretta designs which followed generally their smaller forerunners; the 199 c.c. Zündapp Bella with four-speed gearbox and electric starter; the 197 c.c. Maico Mobie; the 194 c.c. Durkopp Diana with electric starter; the 197 Bond P4 with a Villiers (59 mm. by 72 mm.) which developed 8·5 h.p. at 5000 r.p.m.; the 175 c.c. Phoenix which was also fitted with the same Villiers engine; and the N.S.U. Prima of 174 c.c. capacity (62 mm. by 57·5 mm.) which had a compact engine and gear unit. The cylinder axis of the N.S.U. Prima was set across the centre line of the machine, and the 4-speed gearbox and the spiral-bevel final drive formed a pivoted unit which carried the rear wheel; a Bosch starter-generator was fitted. An unusual American design was the 164 c.c. Harley-Davidson Topper model, which had a horizontally-mounted two-stroke engine with a reed induction valve, and automatic transmission by means of vee belt primary drive carried on expanding and contracting pulleys. Examples of designs using four-stroke engines were the 174 c.c. (60 mm. by 61·5 mm.) Heinkel Tourist (Plate 16) whose single-cylinder overhead-valve engine developed 9·2 h.p. at 5500 r.p.m., and was fitted with a 12-volt Bosch starter generator unit. The largest example of this class was the 200 c.c. TWN Contessa design which had a twin-piston two-stroke engine having an output of 10·4 h.p. at 4800 r.p.m. The 198 c.c. (64 mm.

by 62 mm.) Zündapp Bella 204 model developed 12 h.p. at 5400 r.p.m. The newest and most advanced example of this class was the Lambretta SX 200 design which had a single-cylinder two-stroke engine of 198 c.c. capacity (66 mm. by 58 mm.) and developed 11 h.p. at 5500 r.p.m. The drive was by chain, enclosed in a cast-alloy oil bath case, from the engine shaft to a 4-speed and clutch unit driving the rear wheel directly through an extension of the output shaft. The front brake was of the disc type.

The largest and most powerful examples of motor scooters included such designs as those fitted with large single-cylinder two-stroke air-cooled engines such as the German 247 c.c. (67 mm. by 70 mm.) Maicoletta; the unusual and more complex 248 c.c. (54 mm. by 54 mm.) twin-cylinder, horizontally-opposed, two-stroke with reed inlet valves engined Velocette Viceroy which had a primary shaft drive to a 4-speed gearbox with output shaft driving the rear wheel through a pinion and bevel; and the 249 c.c. (56 mm. by 50·6 mm.) Triumph Tigress/B.S.A. Sunbeam fitted with a twin-cylinder four-stroke vertical engine (Fig. 13). These largest engined motor scooters with power outputs of 10 to 12 h.p. at crankshaft speeds of about 5500 r.p.m., were capable of continuous cruising speeds of up to 60 m.p.h. with top speeds some 10 m.p.h. higher; they provided both greater performance and higher carrying capacity, while improved riding comfort was achieved through elaborate hydraulically-damped coil spring front and rear suspension. The weight of about 240 lb. made these largest machines manageable for the normally robust rider, and they were intended for fast long-distance touring with passenger and luggage rather than for town work. The proprietary 249 c.c. Villiers 4T two-stroke twin-cylinder engine and gear unit, which developed 17 h.p. at 6000 r.p.m. and incorporated fan cooling and a 12-volt Siba starter-generator, was used with the British DKR Manx model.

LIGHTWEIGHT MOTOR CYCLES

The miniature or lightweight motorcycle of up to 250 c.c. capacity was, unlike the newly conceived cyclemotor and the motor scooter, of long established form and the models of the 1960's conformed generally to the classic essentials which had been evolved and used during several decades. Nevertheless, new and progressive ideas had modified this classic form in certain important details, chiefly in frame design, so that current motorcycle models acquired a streamlined and lightened appearance. In the smaller categories, design was influenced by cyclemotor practice, particularly in such aspects as the simple spine frame, or the light stiff duplex diamond frame, and the use of the engine crankcase or even the whole engine unit incorporated within the frame to add to its strength and stiffness. More detailed items included such features as more powerful brakes, efficient front and rear suspension of the hydraulically damped spring coil kind, and full electric equipment including in some cases electric starting. The high specific outputs of the engines of this period, of both the single and multi-cylinder two- and four-stroke varieties of small compact form and much increased crankshaft speeds, invested these small machines with performances which made them more popular in comparison with larger motorcycles, and began to replace them for a wide variety of uses. These small engines were also notable for the still higher outputs and crankshaft speeds to which they were capable of being super-tuned with surprising reliability for sustained racing. Examples of these high outputs are noted in the following summary of contemporary designs of this lightweight class of motorcycles.

The smallest class of lightweight motorcycle had 50 c.c. capacity engines, of both the two-stroke and four-stroke kinds, which developed from 2 h.p. to 4·5 h.p. at crankshaft speeds of from 5000 r.p.m. to 9500 r.p.m. The frames were generally of the overhead spine type, with the engine unit attached at the lower end with the cylinder inclined forward or positioned horizontally (Plate 17). A 3- or 4-speed gearbox was fitted, in place of the simpler performance requirements of the cyclemotor, and the rear chain drive was now frequently enclosed in a case. Among the lower-powered examples of this small category with two-stroke engines were the Foster-Ital; the Garelli Monza; the Jawa; the Kreidler J.51; the N.S.U. Quickly-TTK; the Püch Sports; the Suzuki M12; and the Yamaha 50. Designs of this small size having four-stroke engines included the 50 c.c. (39 mm. by 42 mm.) Ariel Pixie (Fig. 14) which had an overhead-valve engine and developed 3·8 h.p. at 9000 r.p.m.; and the 63 c.c. (44 mm. by 41 mm.) Honda S65 which had an overhead-camshaft engine and a power output of 6·2 h.p. at 10,000 r.p.m. This latter commercial design directly reflected the super-tuned outputs obtained with a variety of specialized racing engines, such as the 50 c.c. (33 mm. by 29 mm.) twin-cylinder Honda engine having four valves per cylinder which developed 12 h.p. at crankshaft speeds of up to 19,000 r.p.m., and which gave a road speed of more than 90 m.p.h.; and the twin-cylinder 50 c.c. Suzuki and Yamaha two-stroke engines having like performances. A later Suzuki design having three water-cooled cylinders and disc inlet valves was reported as being capable of developing 20 h.p. at 20,000 r.p.m.

The next highest class of lightweight motorcycles of up to 125 c.c. engine capacity, were necessarily somewhat heavier and more complex, yet they possessed the same characteristics of lightness and simple progressive design inherent in the smaller machines. This larger category was represented in the two-stroke class by such designs as the Suzuki 120 model whose 118 c.c. (52 mm. by 56 mm.) single-cylinder engine developed 10 h.p. at 7000 r.p.m. with a compression ratio of 7·2 to 1. The lubrication of this engine was effected by an engine-driven metering pump which fed oil proportioned to load and speed direct to the engine main bearings and big end, and thence by splash and mist to the cylinder wall and piston. The drive to the rear wheel was taken through a 4-speed constant-mesh gearbox, a multi-plate clutch running in oil and a rear chain enclosed in a case. The maximum speed was 68 m.p.h. and the fuel consumption was 150 m.p.g. at 40 m.p.h. The light welded steel overhung spine frame contributed to the low overall weight of this machine of 189 lb. The Suzuki Sports 80 was a scaled-down version of the 120 model, with an 80 c.c. capacity engine (45 mm. by 50 mm.) which developed 7·3 h.p. at 7000 r.p.m. and had a maximum speed of nearly 60 m.p.h. The Yamaha 123 design had similar characteristics. Other single-cylinder two-stroke European models of this category included the 110 c.c. Moto Guzzi; the 124 c.c. Capriolo; the 124 c.c. MZ; the 125 c.c. Bultaco; the 125 c.c. Püch; and the 125 c.c. Montesa. Twin-cylinder two-stroke designs included the

125 c.c. Suzuki whose engine had rotary-disc inlet valves and developed 26 h.p. at 12,500 r.p.m.; and the 97 c.c. Yamaha (38 mm. by 43 mm.) (Fig. 15) which had rotary-inlet valves and developed 9·5 h.p. at 8500 r.p.m.

Examples of four-stroke engined models included the 75 c.c. (47·6 mm. by 42 mm.) B.S.A. Beagle (Fig. 16) which had a neat overhead-valve engine and gear in unit construction; the Honda S90 which had an overhead camshaft four-valve single-cylinder engine of 98·5 c.c. capacity (50 mm. by 46·6 mm.) which developed 8 h.p. at 9000 r.p.m., and had a light overhung spine frame; and the 123 c.c. Gilera with push-rod overhead valves.

Specially tuned and more complex engines having either single cylinders, or small-capacity multi-cylinders to permit a substantial increase of crankshaft speeds intended for racing, were produced successfully in a variety of designs. Among the two-stroke engined single cylinder machines were the German MZ 124 c.c. (54 mm. by 54 mm.) which developed 25 h.p. at 10,800 r.p.m., and the earlier Spanish 125 c.c. (51·5 mm. by 60 mm.) Montesa which developed 18 h.p. at 8000 r.p.m. Examples of twin-cylinder racing engines were the 125 c.c. (43 mm. by 43 mm.) Italian Mondial which developed 30 h.p. at 14,000 r.p.m.; and the 125 c.c. (43 mm. by 43 mm.) Yamaha which developed 25 h.p. at 12,500 r.p.m. Design features such as compression ratios of up to 14 to 1, rotary-disc inlet valves, light-alloy cylinders with chromium-plated bores, and even four-cylinder two-stroke arrangements for the small total capacity of 125 c.c. were adopted by the Japanese firms of Suzuki, Yamaha and Kawasaki to achieve even greater power.

Single-cylinder four-stroke engines of 125 c.c. category included such examples as the 124 c.c. (55 mm. by 52 mm.) Capriolo with an overhead-camshaft engine which developed 7·2 h.p. at 5000 r.p.m.; the push-rod overhead valve 124 c.c. (53 mm. by 56 mm.) Bianchi Bernina; and the later 123 c.c. (56 mm. by 50 mm.) Gilera with a higher compression ratio of 8·7 to 1 which gave this machine a maximum speed of 70 m.p.h. and a cruising fuel consumption of 120 m.p.g. An example of a twin-cylinder four-stroke engine was the 124 c.c. (44 mm. by 41 mm.) Honda CB92 model, which had a maximum crankshaft speed of 10,500 r.p.m. and a 10 to 1 compression ratio. This was the forerunner of the 154 c.c. (49 mm. by 44 mm.) Honda C95 model which, with the slightly lower compression ratio of 8·3 to 1 developed 13·5 h.p. at 9500 r.p.m.; and the Honda CB160 model with a 161 c.c. (50 mm. by 41 mm.) engine which developed 16·5 h.p. at 10,000 r.p.m. Both these Honda twin-cylinder designs had overhead-camshaft operated valves and electric starting, 4-speed gearboxes and multi-plate clutches running in oil. Highly-tuned racing engines of 125 c.c. capacity included the 125 c.c. (55·25 mm. by 52 mm.) Ducati single-cylinder engine equipped with positive-action desmodromic valve operation; this engine developed 19 h.p. at 12,500 r.p.m. compared with 16 h.p. at 12,000 r.p.m. which this same engine developed when equipped with double-overhead camshaft valve operation. The 125 c.c. CZ twin-cylinder double-overhead camshaft engine developed 24 h.p. at 15,600 r.p.m. The more complicated double-camshaft four-valve per cylinder 125 c.c. four-cylinder Honda racing engine developed 27·5 h.p. at 18,000 r.p.m. The need to increase the crankshaft speed still more to obtain more power introduced an experimental 125 c.c. five-cylinder Honda with still smaller cylinders.

A considerable variety of production designs with engine capacities up to 200 c.c. were generally available for sale to the public. The two-stroke varieties included the single-cylinder 197 c.c. James Captain; the 175 c.c. B.S.A. Bantam; the 196 c.c. Bultaco; the 197 c.c. D.K.W.; and the twin-cylinder 199 c.c. Ariel Arrow; the 150 c.c. Püch Alpine and the 150 c.c. E.M.C. The 199 c.c. Triumph Tiger Cub exemplified the single-cylinder four-stroke design; while twin-cylinder models included the Velocette water-cooled LE 200 and the air-cooled Valiant, while this basic design was used for the still later more elaborate Vogue, which was adopted for light police patrol work.

In the next range of engine capacity up to 250 c.c. there was a large production of a variety of designs for both touring and racing. The single-cylinder two-stroke category included the 249 c.c. (65 mm. by 75 mm.) Jawa; the 250 c.c. Bultaco Metralla which developed 27 h.p. at 8000 r.p.m.; the 249 c.c. Greeves; and the Spanish Ossa of 230 c.c. capacity, which developed 22 h.p. with a compression ratio of 10 to 1. The twin-cylinder two-stroke category included the 249 c.c. (50 mm. by 63·5 mm.) Villiers engined DMW Sports; the larger 247 c.c. (54 mm. by 54 mm.) Ariel Arrow (Fig. 17); the 247 c.c. Suzuki T20 (54 mm. by 54 mm.) which developed 29 h.p. at 7500 r.p.m. and was fitted with twin carburettors, and positive oil feed; the 246 c.c. (56 mm. by 50 mm.) Yamaha (Fig. 18) which developed 25 h.p. at 7500 r.p.m. and was capable of running up to 10,000 r.p.m. in the lower gears; it also had a compact 5-speed gear assembly incorporated neatly in unit with the engine crankcase; the 250 c.c. Kawasaki which developed 31 h.p. at 8000 r.p.m.; and the 248 c.c. (45 mm. by 78 mm.) Püch twin-piston design which had long been a feature of this manufacturer.

The single-cylinder four-stroke engined motorcycles of about 250 c.c. numbered a quantity of long-established designs brought up to date and given improved performances, such as the 246 c.c. (66 mm. by 72 mm.) Aermacchi; the 249 c.c. (67 mm. by 70 mm.) B.S.A.; the 204 c.c. (67 mm. by 57·8 mm.) Ducati whose overhead-camshaft engine developed 18 h.p. at 7,500 r.p.m.; the 248 c.c. (70 mm. by 65 mm.) A.J.S. (Plate 18) and Matchless; the 235 c.c. (68 mm. by 64 mm.) Moto-Guzzi; the 247 c.c. (69 mm. by 66 mm.) N.S.U. Super Max; and the 248 c.c. (70 mm. by 64 mm.) Royal Enfield. All these standard production models were of normal overhead-valve push-rod design. The more complex twin-cylinder production designs included the 249 c.c. (60 mm. by 44 mm.) Norton Jubilee (Plate 18); and the 247 c.c. (54 mm. by 54 mm.) Honda with overhead-twin-camshaft operated valves; the 245 c.c. (68 mm. by 68 mm.) B.M.W. which had the distinction of final shaft drive.

The engine designs in this 250 c.c. category produced during this period became increasingly complex and progressive and the resulting power outputs and road speeds more than ever outstanding. The high performance of the two-stroke engines of this period was primarily due to the system evolved by the German MZ concern for their 125 c.c. and 250 c.c. engines with water cooling, single and even twin-rotary inlet valves to increase charge efficiency and establish asymmetrical inlet timing, three transfer ports, half-moon 'squish' (turbulence) combustion chamber, and resonant rear exhaust system. This two-stroke system was adopted by the chief Japanese two-stroke engine manufacturers Suzuki and Yamaha with outstanding success. The high performance four-stroke engines achieved their high specific outputs primarily by a very substantial increase in crankshaft speeds with good volumetric efficiency; this was obtained by the use of multi-small-bore cylinders each with four valves operated by twin camshafts as illustrated, for instance, by such designs as the Honda 125 c.c. four-cylinder and 250 c.c. four-cylinder and six-cylinder (Plate 19) designs. The practical application of these fundamental principles for both two and four-stroke engines may be seen in the brief description which follows of the more important racing models used with remarkable success in the later 1960's.

The earlier two-stroke 247 c.c. (54 mm. by 54 mm.) air-cooled MZ racing engines designed on the principles stated above developed 43 h.p. at 12,000 r.p.m., and a later water-cooled version developed 48 h.p. at 10,500 r.p.m. Both Suzuki and Yamaha followed the MZ principles with their 250 c.c. twin-cylinder designs and later produced the more complex duplex arrangement of four-cylinder square water-cooled units (Plate 19) using smaller cylinder bores to achieve still higher crankshaft speeds.

The 250 c.c. twin-cylinder four-stroke Jawa design was an example of the earlier and simpler form of racing engine in this class, having two valves per cylinder operated by twin-overhead-camshafts. Later more progressive designs were exemplified by the 250 c.c. three-cylinder MV Agusta which had four valves per cylinder operated by twin-overhead camshafts; the Benelli four-cylinder design; and the 250 c.c. Ducati design with desmodromic valve-operating gear (Fig. 19). The Honda concern made further daring progress with their policy of increasing the number of cylinders and reducing their individual capacity to achieve higher crankshaft speeds with their 250 c.c. (44 mm. by 41 mm.) four-cylinder engine design which developed 46 h.p. at 13,800 r.p.m.; and later by their 250 c.c. six-cylinder design (Plate 19), which developed 51 h.p. at 16,000 r.p.m.

Even more progress in engine design may be possible through the adoption of even larger numbers of smaller cylinders and still higher crankshaft speeds. It is probable that eventually 40 h.p. may be obtained from a 125 c.c. engine and 75 h.p. from a 250 c.c. engine.

A specialized class of high-power single-cylinder two-stroke engines such as the 246 c.c. Villiers 36A design, which developed 31 h.p. at 7500 r.p.m., and the 246 c.c. Greeves which developed 33 h.p., were intended for trials, scrambles and moto-cross sporting events.

50 c.c. Raleigh Wisp cycle-motor: 1967

49 c.c. Honda 50 cycle-motor: 1966

49 c.c. Suzuki 50 cycle-motor: 1966

Plate 16

88 c.c. Vespa 90 motor-scooter: 1966

148 c.c. Lambretta SX 150 motor-scooter: 1966

174 c.c. Heinkel Tourist motor-scooter: 1960

152 c.c. Ducati Monza Junior motorcycle: 1965

97 c.c. Yamaha motorcycle: 1966

79 c.c. Suzuki 80 motorcycle: 1966

49 c.c. Kreidler Florette Super motorcycle: 1965

Plate 18

305 c.c. Honda 305 motorcycle: 1966

248 c.c. A.J.S. motorcycle: 1958

249 c.c. Norton Jubilee motorcycle: 1952

Motorcycles CHAPTER 9
Larger Motorcycles: after 1956

The popular success of the smaller categories of motorcycle during the 1960's, outlined in the preceding chapter, which was due to their essential economy, handiness, high comparative performance and sure reliability, had the effect of shifting the motorcycle utility scale lower in terms of size, weight and engine capacity so that a preponderance of motorcycle users opted for the smaller machine to the partial neglect of the larger. In addition, motorcyclists now included a large proportion of youthful as well as older riders of both sexes whose essential requirement was a selection of handy runabouts for town and country work and who naturally preferred the smaller, more economical mount to the larger and heavier more conventional machine. Another social aspect which made the larger motorcycle now less sought after was a more affluent lower class who could afford at least a second-hand four-seat saloon minicar at the price of, or even less than, a new 500 c.c. motorcycle. This same influence seriously lessened the appeal of the coach-built family sidecar outfit which previously had been a boon to lower salaried road users. Because of these new economic conditions, while manufacturers continued to design and produce a full range of larger motorcyles during this period, sales of them declined in favour of the smaller forms of motorcycle, and the demand for the sidecar declined even more seriously with indications that it might eventually disappear except for special uses, one of which, ironically enough, being as carrier tenders for attachment to motor scooters. By the end of the 1960's, therefore, these social changes had divided motorcycles into two broad categories: the lightweight motorcycle had become essentially the utility machine for both short and long distance runs, while the larger motorcycle, except when used for special work such as police patrols, remained in general the mount of the enthusiast.

Technical development still proceeded with the larger machines, and their specifications benefited generally from the progress which was being achieved with the new fast developing forms of light motorcycle. This progress with the larger machines in general, and their engine and gear units in particular, was pursued especially with the larger capacity racing engines which broadly followed in scaled-up form the technical policies which were proving so successful with the smaller machines. Nevertheless, the larger production motorcycle design tended to retain its original conservative style; the older, more established forms continued in production and there was little tendency for design to break new ground as had happened in the lightweight field during the past two decades. Advanced design forms were confined to specialized racing models having engine capacities up to 500 c.c., and a very few production models which were derived from this advanced racing technique.

350 C.C. MOTORCYCLES

The single-cylinder category of 350 c.c. capacity motorcycles, which originated early in the century as a side-valve engined machine and later developed into a push-rod operated overhead-valve engined machine of enhanced power and performance, had long been one of the standard types of medium power motorcycles which continued in steady production with further improvements but without modifying its basic conservative layout. The overhead-valve engines had long strokes at first, but with current engine development demanding higher crankshaft speeds the stroke was shortened until oversquare cylinder dimensions became standard, as will be apparent from the selection of designs mentioned in this chapter. The specification included a 4 and in some cases a 5-speed touring ratio gear train in unit with the engine crankcase, although separate gearboxes were still used in a diminishing number of cases; a compression ratio of about 7 to 1; and a power output of about 20 h.p. at 6500 r.p.m. Sports versions of the same models with a compression ratio of 9 to 1 had increased outputs of about 28 h.p. at 8000 r.p.m. Telescopic spring front forks and coil-spring and hydraulically-damped rear suspension units supporting rear wheel stays were general features of the frame construction. This category of single-cylinder motorcycle of medium capacity was a practical general purpose mount of good average performance and extreme reliability that bridged the small and large categories of motorcycle of this period, and for that reason was more generally popular and in larger demand than the larger versions. Current designs of this category were mostly of British origin and included the 348 c.c. (72 mm. by 85·5 mm.) A.J.S. model 16; the 343 c.c. (79 mm. by 70 mm.) B.S.A. model B40; the 348 c.c. (72 mm. by 85·5 mm.) Matchless model G3; the 348 c.c. (72 mm. by 88 mm.) Norton model 50; the 348 c.c. (71 mm. by 88 mm.) Panther model 75; the 346 c.c. (70 mm. by 90 mm.) Royal Enfield Bullet; and the 349 c.c. (72 mm. by 86 mm.) Velocette Viper. There was little interest from Continental manufacturers in this size and type of standard motorcycle, their effort being concentrated rather upon the wide variety of progressive types described in Chapter 8. This fact emphasizes the obsolescence of the 350 c.c. single-cylinder formula.

The more complex and up-to-date 350 c.c. twin-cylinder designs had a greater flexibility of performance and handling quality and were more favoured by Continental as well as British manufacturers, both for standard touring purposes as well as for racing. Examples of this four-stroke twin-cylinder category included the advanced 305 c.c. (60 mm. by 54 mm.) Honda C77 model, (Plate 18), which had twin-overhead camshaft-operated valves and developed 23 h.p. at 7500 r.p.m. as a standard model, and 28·5 h.p. at 9000 r.p.m. with a compression ratio of 9·5 to 1 as a sports model; the 350 c.c. Parilla GT model; the 349 c.c. (63 mm. by 56 mm.) Norton Navigator model; and the 348 c.c. (56·25 mm. by 65·5 mm.) Triumph Tiger 90 model which, with a compression ratio of 9 to 1, developed 27 h.p. at 7500 r.p.m.

Of the two-stroke designs of this 350 c.c. category, the proprietary 323 c.c. (57 mm. by 63·5 mm.) Villiers Mark 3T twin-cylinder engine and 4-speed gearbox unit with fan cooling and Siba Dynastart electric starting and reverse switching was used in some British models, including the D.M.W. Dolomite model; the Greeves 32DC model; the Panther 45 model; and the Cotton Messenger model. Other individual designs of twin-cylinder two-stroke engines were the 328 c.c. (58 mm. by 62 mm.) Excelsior Talisman S10 model ; and the 344 c.c. (58 mm. by 65 mm.) Jawa Senior model.

A revival of the Scott name, which pioneered the twin-cylinder two-stroke motorcycle engine before 1914, appeared during this period in the form of a twin-cylinder air-cooled design (Fig. 20) of 344 c.c. (60·3 mm. by 60·3 mm.) capacity which, with a 10·5 to 1 compression ratio had a maximum crankshaft speed of 8000 r.p.m. This new high-performance Scott design retained some of the features of the early Scott models such as parallel twin cylinders, a built-up crankshaft with 180 degree throws and mounted on roller bearings, and a central flywheel to which the two separate crankshafts were joined by keyed tapers and a central tension bolt. The crankpins were interference fits in their respective webs, permitting the use of caged needle roller big-end bearings. Another British design of this kind was the 349 c.c. (63 mm. by 56 mm.) R.C.A., which developed 21 h.p. at 5200 r.p.m. Continental proprietary engine manufacturers who produced two-stroke engines of this kind included the large German concern of Sachs. These engines were sometimes fitted with a motor-generator dynamotor type of flywheel unit which could include reverse switching to utilize the two-stroke engine's ability to run in either direction with a re-adjustment of the ignition timing; this provision made these engine units suitable not only for use with ultra-light cars but also for sidecar combinations.

Two high-performance Japanese models in this 350 c.c. category of two-stroke engined production models were the twin-cylinder 338 c.c. (62 mm. by 56 mm.) Kawasaki A7 design, which had disc rotary inlet valves, a 5-speed gearbox, and a lubrication system known as Injectolube, by means of which oil was delivered to the induction tract and the main and big-end bearings by a throttle-controlled pump. This machine weighed only 329 lb. and, with an output of 40·5 h.p., which was developed at 7500 r.p.m., was capable of a speed in excess of 100 m.p.h. The 348 c.c. (61 mm. by 59·6 mm.) Yamaha R1 twin-cylinder model, although somewhat heavier at 385 lb., was also capable of maximum speeds of more than 100 m.p.h. Another class of high-power two-stroke engined production models were those made for use in trials, scrambles and moto-cross events, such as the Greeves and CZ designs, whose 360 c.c. single-cylinder two-stroke engines developed some 33 h.p. at 6200 r.p.m. These specialized motorcycle designs were of simple and sturdy construction, devoid of unwanted complication and weight and able to withstand the abnormal ground shocks that this form of motorcycling sport entails.

In the 350 c.c. category of racing motorcycle, intense development continued on the lines of the new and progressive engine techniques so successfully applied to the smaller racing capacities, as already described. It is to be noticed that this racing progress, as a forerunner to still higher and more efficient commercial production, was now essentially of foreign origin and inspiration, notably from Germany, Italy and Japan, and these up-to-date multi-cylinder designs seriously outclassed the obsolescent British single and twin-cylinder engined machines which still appeared in competition road and track race events. While British motorcycle production models of conventional design continued to retain their essential thoroughness and quality which had always characterized them, the bold innovations of this period which foreign manufacturers were now testing under sustained and arduous operational conditions, were applied to their production models with a like high manufacturing quality and reliability in operation. This will be apparent from the specifications and performance figures included in this and the preceding chapter, while in comparison the British specifications and performance figures were obviously of a more conservative nature.

In considering the increased outputs of these 350 c.c. racing engines in comparison with the smaller categories already described, it is to be noted that their road performances did not rise in linear proportion. This was due to some extent to the twisting and hilly conditions of such courses as the Isle of Man Tourist Trophy course imposing a limitation on possible full power utilization. On the other hand, the example of a 297 c.c. Honda six-cylinder machine winning the 350 c.c. Junior Tourist Trophy Race in 1967 at an average speed of 107 m.p.h., would indicate that the smaller and lighter machine with the higher power-weight ratio can apply its available power more efficiently than heavier and more powerful machines. The indication is apparent that there is a watershed of practical efficiency which now exists between 250 c.c. capacity machines on the one hand and larger machines on the other, which had been established by the high specific outputs now being obtained from small capacity engines. This emphasizes the clear distinction between the two broad categories of machines dealt with respectively in Chapters 8 and 9.

The 348 c.c. (65 mm. by 52·5 mm.) Bianchi twin-cylinder two-valve design which developed 48 h.p. at 10,600 r.p.m., was an earlier example of this racing formula. The primary drive was taken from the centre of the built-up crankshaft to minimize torsional fluctuations. The later 349 c.c. (46 mm. by 52·6 mm.) Jawa with two-valve cylinder heads developed 50 h.p. at 10,600 r.p.m.; while the four-valve version of the same design had its crankshaft speed increased to 11,400 r.p.m. and its maximum power to 55 h.p. The earlier four-cylinder Gilera designs such as the 349 c.c. (46 mm. by 52·6 mm.) with four-valve cylinder heads which developed 50 h.p. at 11,000 r.p.m., and the 349 c.c. (49 mm. by 45 mm.) which developed 52 h.p. at the higher crankshaft speed of 12,000 r.p.m.; were significantly surpassed by such later designs as the 297 c.c. Honda which developed 65 h.p. at 16,000 r.p.m., and which won the 1967 Junior Tourist Trophy 350 c.c. Race at the average speed of 107·73 m.p.h.; and the 350 c.c. MV Agusta which had a three-cylinder engine and developed maximum horse-power at 12,500 r.p.m.; and the still later 350 c.c. four-cylinder Benelli design. After the success of the 297 c.c. Honda in the 1967 Junior Tourist Trophy Race already mentioned, a 350 c.c. six-cylinder Honda design was projected for the next season.

Although two-stroke four-cylinder Suzuki and Yamaha engine designs had been developed and used in first-class racing in the 250 c.c. class, complex engines of this type were not yet developed sufficiently to be wholly reliable and further experimental work continued with this capacity of racing engine. A four-cylinder two-stroke arrangement is complicated by the necessity of crankcase induction, and still further by the addition of disc rotary induction valves which are necessary to achieve the still higher power outputs required. These requirements result in a bulky and complex power unit made more so by the necessity of water-cooling cylinder jackets and a radiator. Other two-stroke designs of a similar high power nature were also being evolved in Europe such as the four-cylinder 350 c.c. (48 mm. by 47 mm.) Jawa, which developed 60 h.p. at 13,500 r.p.m.; and the two-cylinder 347 c.c. (80 mm. by 69 mm.) CZ design, which developed 46 h.p. at 10,000 r.p.m. Another four-cylinder two-stroke racing engine was the 350 c.c. Russian Vostok design.

500 C.C. MOTORCYCLES

The 500 c.c. single-cylinder four-stroke engined motorcycle still continued in conservative production, although it was now quite outmoded for high-performance racing purposes, and had perhaps reached its peak for racing more than a decade earlier. These standard production push-rod engines were rated at some 30 h.p. at 6000 r.p.m. at the moderate compression ratio of 7·5 to 1. Typical models included the 498 c.c. (82·5 mm. by 93 mm.) A.J.S. Model 18; the 499 c.c. (85 mm. by 88 mm.) B.S.A. model B34; the 498 c.c. (82.5 mm. by 93 mm.) Matchless model G80; the 490 c.c. (79 mm. by 100 mm.) Norton model ES2; the 499 c.c. (86 mm. by 85·6 mm.) Norton model 30M, which had overhead-camshaft operated valves; and the 499 c.c. (84 mm. by 90 mm.) Royal Enfield Bullett.

Examples of the 500 c.c. capacity twin-cylinder production motorcycle of this period had engines of the parallel twin-cylinder air-cooled arrangement, with the exception of the horizontally-opposed 490 c.c. (68 mm. by 68 mm.) B.M.W. model R50. The parallel twin-cylinder designs included the 498 c.c. (66 mm. by 72·8 mm.) A.J.S. model 20 (Fig. 24); the 497 c.c. (66 mm. by 72·6 mm.) B.S.A. model A7 (Fig. 24); the 441 c.c (79 mm. by 90 mm.) B.S.A. model B44; the 498 c.c. (66 mm. by 72·8 mm.) Matchless model G9; the 497 c.c. (66 mm. by 72·6 mm.) Norton model 88; and the 496 (70 mm. by 64·5 mm.) Triumph Speed Twin. The 444 c.c. (70 mm. by 57·8 mm.) Honda 450 (Fig. 22) was a more progressive example of this usually conservative class, with its advanced design characteristics such as an 180 degree crankshaft, twin-overhead camshafts and high output of 43 h.p. at 8500 r.p.m. It is to be noted that the vee-twin cylinder engine, formerly so popular for the higher powered motorcycle, had now been abandoned in favour of the more compact forms mentioned above.

A limiting capacity of 350 c.c. existed for two-stroke engines of both the production and racing kinds chiefly because of acute thermal problems, particularly in connection with the four-cylinder two-stroke racing engines of 350 c.c. capacity. A sole example of two-stroke engine design in this larger capacity of 500 c.c. was produced during this period in the 492 c.c. (59 mm. by 62 mm.) Excelsior three-cylinder in-line air-cooled engine, which was developed from the earlier 328 c.c. twin-cylinder engine mentioned in Chapter 7. This engine unit included a Siba Dynastart generator and multi-plate clutch driven from the crankshaft by a duplex chain; it was originally intended for installation in the Berkeley sports car.

In the 500 c.c. racing engine class, the foremost designs were of the four-cylinder four-stroke kind, and continued development and use in first class racing events produced ever increasing specific power outputs and performance and a tendency to multiply the cylinders to six and even eight in the interests of attaining still higher crankshaft speeds and outputs. Stemming from the early Gilera and NSU designs of four-cylinder air-cooled racing engines of the 1950's which developed some 50 h.p., the designs of the later 1960's included such examples as the 493 c.c. (52 mm. by 58 mm.) Gilera which developed 70 h.p. at 10,500 r.p.m.; the 500 c.c. MV Agusta and the 500 c.c. Honda which developed 85 h.p. at 12,500 r.p.m. The later development of the three-cylinder MV Agusta was a successful attempt to simplify these high power units and reduce their bulk and frontal area. Nevertheless, the tendency to increase the number of cylinders was growing, and already a 500 c.c. Moto Guzzi vee eight-cylinder water-cooled unit of compact design had been built in 1957, which had proved to be the most powerful racer of this capacity.

750 C.C. MOTORCYCLES

Of the motorcycle designs with engine capacities up to 750 c.c., the 589 c.c. (87 mm. by 100 mm.) Panther model 100 and the 645 c.c. (88 mm. by 106 mm.) Panther model 120 were the sole representatives of single-cylinder design in this category, having been successfully continued in production in basically similar form over many years, chiefly as sidecar machines. The parallel twin-cylinders production models included the 646 c.c. (72 mm. by 79·3 mm). A.J.S. model 31 (Plate 21); the 646 c.c. (70 mm. by 84 mm.) B.S.A. model A10; the 646 c.c. (72 mm. by 79·3 mm.) Matchless model G12; the 597 c.c. (68 mm. by 82 mm.) Norton model 99; the 745 c.c. (73 mm. by 89 mm.) Norton Atlas (Plate 20); the 692 c.c. (70 mm. by 90 mm.) Royal Enfield Super Meteor; and the 649 c.c. (71 mm. by 82 mm.) Triumph model T110. Two production roadster models which derived directly from racing designs and experience were the 591 c.c. (56 mm. by 60 mm.) MV Agusta equipped with a four-cylinder in-line engine which developed 52 h.p. at 9000 r.p.m. and which was capable of a maximum speed of 113 m.p.h.; it was fitted with twin mechanically-operated disc front brakes to control this high speed on the open road. The second production roadster model, although derived from racing practice, was of a radically different layout than that currently adopted for racing; this was the 704 c.c. (80 mm. by 70 mm.) Moto-Guzzi which had a transversely-mounted vee twin-cylinder engine having light-alloy cylinders with unlined chromium-plated bores, an electric starter, a car-type clutch, a four-speed gearbox and final shaft drive.

1200 C.C. MOTORCYCLES

The largest forms of motorcycle had now become quite rare because of the demand being concentrated upon the smaller but relatively high powered and more economical designs, although such designs as the 997 c.c. Ariel Square Four (Plate 22) and the 900 c.c. Harley-Davidson vee twin-cylinder models still continued in small production.

The largest category of motorcycle was represented by one or two current models from the United States of America and still in the traditional form of the large-capacity vee twin-cylinder. This was a formula which American motorcycle manufacturers had made their specialized own in the early years of the century and which was represented by a wide variety of designs already mentioned, particularly in Chapter 4. Only one of these earlier firms survived to this period, namely the Harley-Davidson Company of Milwaukee, Wisconsin, whose Electra Glide and Sportster models to this specification were used largely for mobile police patrol work as well as in a more limited way as ordinary touring mounts.

The Harley-Davidson Electra Glide model (Plate 20) was a large fully-equipped motorcycle weighing 660 lb. Its tubular diamond frame was duplicated to provide the necessary strength and stiffness for so heavy and powerful a machine. The air-cooled vee twin-cylinder push-rod overhead-valve engine of 1200 c.c. capacity (87·5 mm. by 100·5 mm.) developed 66 h.p. at 5600 r.p.m. with a compression ratio of 8 to 1. The ignition was provided by separate coils and contact breakers for each cylinder, the contact breakers being operated by a single cam and the timing interval being arranged by the spacing of the contact breakers. The drive to the rear wheel was taken through an enclosed primary chain, a dry multi-plate clutch, a 4-speed constant-mesh or a 3-speed and reverse constant-mesh gearbox, and an open secondary chain. The reverse gear was intended for sidecar work, a necessary addition with so heavy a combination. Incorporated with the gearbox was either a kick-starter mechanism or an electric starter motor and Bendix pinion which engaged with a gear ring formed on the periphery of the clutch drum. Electric power supply was provided by a direct-current generator, gear-driven from the engine timing gear train, which supplied current for both starting and lighting purposes. The brakes were of the drum type; the front hand brake was cable operated and the rear foot brake was hydraulically operated, the latter feature being a significant addition to so heavy a machine.

CHAPTER 10
Accessories and Ancillaries: from 1956

The individual and specialized components and assemblies which go to make up the complete motorcycle naturally benefited from the extensive general progress which was being made in parallel techniques, notably those related to motorcar design and production. Substantially higher specific outputs were now being obtained from engines having reliable increased performance and longer operational life; more highly specialized gears and transmissions provided greater operational flexibility, especially for the smaller, lighter mounts; novel frame designs provided individual suitability for different categories of machines, together with operational stability and manufacturing economy and standardization; improved drum type brakes of increased friction area and an experimental approach to disc brakes were adopted; and progress in tyre construction and tread pattern aided performance under varying operational conditions as well as providing improved immunity from non-adhesion on different road surface conditions. All this technical progress was rigorously proved in prolonged individual testing and in the arduous operational conditions of sustained road racing before it was applied with confidence to production models.

ENGINES

The technical progress made with motorcycle engines during the decade of the 1960's was notable, not only for the substantial increase in specific outputs of both production and specialised racing designs, but also for the wide variety of design forms of both the two and four-stroke categories of capacities from 50 c.c. to 1200 c.c. which were produced successfully and established commercially. The high specific power outputs now being reliably obtained, together with the ever present desire for economy in both first cost and operation, resulted in a substantial proportion of production designs to be of the lower cylinder capacities of up to 250 c.c.; while models with higher cylinder capacities were in less demand. While the simpler engine designs, in particular those of the two-stroke variety, altered little in external appearance, they nevertheless gained in intrinsic efficiency from the ever increasing knowledge of cylinder head design and efficient gas flow and the scientific promotion of turbulence ('swirl') to improve volumetric efficiency and completeness of combustion; of carburation, with an experimental approach to solid fuel injection; of tuned exhaust flow, especially in the case of two-stroke engines, as an aid to fresh gas charging; the adoption of over-square cylinder dimensions and four valves per cylinder operated by overhead camshafts to permit the safe use of still higher crankshaft speeds; and of progressively more suitable materials which would safely permit the higher mechanical and thermal stresses which this vigorous progress entailed. The bore and stroke dimensions of specific designs are given in Chapters 8 and 9 to illustrate this modern trend towards square and oversquare cylinder dimensions to maintain piston speeds within safe bounds as the crankshaft speeds increased.

The conservative single and twin-cylinder four-stroke engines, evolved from past decades of steady development, retained their original characteristics of sturdy ultra-reliable designs having two push-rod operated valves per cylinder, detachable light-alloy cylinder heads with steel valve seat insets, crankshafts incorporating internal flywheels with pressure lubricated plain big-end and main bearings or, more rarely for this class of engine, a built-up crankshaft with needle-roller big-end and ball-and-roller main bearings. Although the separate gear-box was still used to some extent, there was a growing tendency to incorporate the four- or five-speed gear train compactly in an extension of the crankcase with direct gear primary drive between engine crankshaft and gear input shaft. The majority of twin-cylinder engines, which were now almost always of the parallel disposition, to the almost total exclusion of the formerly popular vee-twin arrangement and with only very few examples of the horizontally-opposed arrangement, had 360 degree crankshafts which caused both pistons to rise and fall in unison. This arrangement gives equal firing periods, but has the imperfect balance characteristics of the single cylinder engine accentuated by the double weight of the two pistons and connecting rods. These conservative designs, most of which were of British origin, developed some 60 h.p. per litre at about 5000 r.p.m. with a compression ratio of 7·5 to 1; sports versions with a compression ratio of 9 to 1, and with over-square cylinder dimensions and crankshaft speeds increased to 7000 r.p.m., developed some 80 h.p. per litre capacity.

The more progressive European designers in general and the late comer Japanese designers in particular were not bound by these old established conservative standards. Their new conceptions considerably transcended these older standards in their production designs and very considerably in their racing designs, which latter served effectively as the experimental testing ground for the former products. Not only were progressive push-rod engine designs with advanced characteristics produced, but also still more advanced designs incorporating overhead camshaft-operated valves of even the smallest capacity — 50 c.c. — were established in reliable commercial production; while, for racing purposes, four valves per cylinder operated by twin-overhead camshafts as in the case of Honda designs, and even by positively operated (desmodromic) valves as in the case of the Ducati designs (Fig. 19) were used. Honda parallel twin-cylinder engine designs were distinctive in having 180 degree crankshafts which, while giving uneven firing periods, resulted in better balance characteristics and of smoother running, particularly at the very high crankshaft speeds — up to 10,000 r.p.m. — used by Honda for their production designs. These more advanced unconventional foreign designs, by virtue of these bold and progressive design features, had considerably higher specific outputs than the conventional designs mentioned above. In production models, power outputs averaged some 100 h.p. at crankshaft speeds of between 8000 and 10,000 r.p.m. with compression ratios up to 10 to 1. The special racing designs from which these high output production designs were evolved, particularly of the four and six-cylinder varieties, adopted very small cylinders — 25 c.c. to 50 c.c. capacities — of over-square dimensions and having four valves per cylinder to permit the safe use of crankshaft speeds approaching 20,000 r.p.m, and to avoid excessive thermal and inertia problems and penalties. With these devices, specific outputs of

230 h.p. per litre were developed in the case of the smallest engines and 170 h.p. per litre in the case of the larger engines, at crankshaft speeds of between 15,000 and 20,000 r.p.m. with, it may be emphasized, remarkable racing reliability which was imparted to their lower rated production designs.

Two-stroke engines, while retaining essentially their basic simplicity of only three moving parts per cylinder unit, also achieved a very high degree of development which produced similar high specific outputs at high crankshaft speeds. This advance was pioneered generally by the German MZ concern, which specialized in efficient two-stroke engine design, and was followed by the Japanese Suzuki and Yamaha concerns with equal success for both their racing and production designs. The features of this advance in two-stroke design included single and even twin-rotary inlet valves per cylinder to provide asymmetrical inlet timing to the crankcase as a means of increasing charging efficiency compared with the symetrical inlet timing of the piston-controlled inlet port; three transfer ports giving scientifically directed gas streams; half-moon turbulence ('squish') combustion chamber shape; and resonant rear exhaust system which materially assisted both exhaust evacuation and inlet charging operations by using the phased inertia energy of the exhaust efflux. In addition, the successful use of light-alloy cylinders having unlined chromium-plated bores provided improved heat transfer and cooling conditions to meet the higher outputs now being developed; also, water cooling was employed on some of the complex multi-cylinder designs specially produced for racing. Among the latter were designs ranging from twin-cylinder units of only 50 c.c. total capacity, which developed some 12 h.p. at crankshaft speeds approaching 20,000 r.p.m., to 250 c.c. capacity four-cylinder water-cooled designs developing 60 h.p. at 15,000 r.p.m. A representative 250 c.c. single-cylinder design for commercial production during their period was capable of developing some 25 h.p. at 7500 r.p.m.

The following comparative table of actual power outputs of modern motorcycle racing engines of from 50 c.c. to 500 c.c. capacities summarizes the state of development achieved by the 1960's.

Make	Capacity in c.c.	Type	bhp at rpm	bhp per litre	Piston speed in ft/min.
Honda	50	ohc twin	11·5/19,000	230	3,950
Suzuki	50	ts single	11/12,000	220	3,200
Honda	125	ohc four	27·5/18,000	220	4,000
Suzuki	250	ts four	54/12,000	215	3,500
Honda	250	ohc six	51/16,000	205	3,900
MZ	250	ts twin	50/10,500	200	4,000
Yamaha	250	ts twin	50/11,000	200	4,000
CZ	125	ohc twin	24/15,600	190	4,050
Honda	250	ohc four	46/13,800	185	3,900
Bultaco	200	ts single	30/9,500	150	4,050
Honda	500	ohc four	85/11,000	170	4,100
MV	500	ohc four	70/10,500	140	4,050
Villiers	250	ts single	31/7,500	125	4,000
Greeves	250	ts single	30/7,500	122	4,000
Norton	500	ohc single	52/7,000	104	4,050

Plate 19

250 c.c. Yamaha 4-cylinder engined motorcycle: 1966

250 c.c. Honda 6-cylinder engined motorcycle: 1966

Plate 20

745 c.c. Norton Atlas motorcycle: 1967

1,200 c.c. Harley-Davidson Electra-Glide motorcycle: 1967

646 c.c. A.J.S. motorcycle and Watsonian Monza sidecar: 1967

Plate 22

Ariel 'Square Four' motorcycle: 1959

Carburettors

Carburettor design for motorcycle engines was generally standardised in most main features with minor differences in detail design. The main essentials were a float chamber and float mechanism controlling the critical level of the petrol at the main jet; main and slow-running jets; throttle slide with taper needle which varied the main jet aperture with respect to throttle opening; and air slide to control mixture proportion. Among the chief manufacturers were the British Amal and Villiers; the Italian Dell'Orto; the German Bins; and the Japanese Kei-Hin designs. Latest designs, such as the Amal Monoblock, having a float chamber concentric with the air intake and main jet, introduced a smaller and more compact trend of design. Metered pressure injection of fuel was applied to a few experimental racing engines.

Lubrication

For four-stroke engines, the normal system of lubricating oil at a pressure of 40 to 50 p.s.i. supplied from and returned to a sump or tank by an engine driven pump had become standard practice for the effective lubrication of main and big-end bearings, and thence by splash and mist to cylinder bores, pistons, small end bearings, timing gears and operating mechanism and various auxiliary drives. In the case of some of the engine designs incorporating the transmission gear train within the crankcase, the lubrication system was common to both.

For two-stroke engines, the simple yet effective petroil system of mixing a small proportion of oil with the petrol was still in almost general use, with the proportion of oil to petrol being substantially reduced to as little as one part in fifty, although an average proportion about one part in thirty was more general. The Suzuki 'Posi-Force', the Yamaha 'Autolube', the Kawasaki 'Injectolube' and the Lambretta 'Lubematic' systems of two-stroke engine lubrication avoided the inconvenience of the petroil system by use of an oil pump which metered the oil supply according to speed and load directly to the main and big-end bearings and thence by splash and mist to the piston and cylinder walls. Replenishment of this system consisted merely in filling an oil tank without the former necessity of mixing the oil with the petrol. These positive systems of two-stroke engine lubrication not only provided more efficient but also more economical lubrication, the average proportion of oil to petrol being about one part to fifty parts.

ELECTRICAL EQUIPMENT

Ignition

The high-tension magneto, as originated by Honold in 1902 and developed in large production by Robert Bosch, was still in general use during this period in improved and considerably lighter and more compact form; it was now used in particular for the larger conservative single and twin-cylinder motorcycles which are dealt with in Chapter 9. Examples of the wound-armature type of magneto were the Lucas K2F and the BTH designs which, although they were similar in principle to the early Bosch design, were distinguished by much smaller and lighter permanent magnets, which had so added to the weight and size of the earlier machines. The Lucas SR1 design was a current example of a rotating magnet or polar inductor type of magneto; because of the absence of armature windings, a higher speed of rotation was possible which made it more suitable for racing purposes.

High-tension coil ignition with an engine-driven distributor incorporating a make-and-break mechanism and drawing its primary current from a generator and battery supply, had the advantage over the magneto in providing a strong high-tension spark at low engine revolutions, thus making for easier starting. It was used on certain four-stroke engined motorcycles, in particular on some of the smaller varieties. The Lucas 18D2 design of distributor unit was a typical example of this device. Another form of distributor and contact breaker assembly was incorporated in the crankcase casting and driven from the half-time camshaft. The essential simplicity of the distributor unit, particularly if used in conjunction with a high-output at low r.p.m. A.C. generator incorporating emergency-start facilities as a precaution against a flat battery, tended to make it preferable to the magneto as an ignition unit.

The flywheel magneto, introduced by Villiers in the early 1920's was used extensively on two-stroke engines where one spark per revolution per cylinder is required. The flywheel magneto unit comprised a fixed backplate, attached to the crankcase, on which were mounted the contact-breaker and both ignition and lighting induction coils, and a flywheel mounted on the crankshaft extension in which were incorporated permanent magnets which revolved past the soft iron laminated armatures on which the induction coils were wound. The Villiers 9E design of flywheel magneto and lighting generator design was typical and was much used for two-stroke engines of both the single and twin-cylinder varieties.

An improved form of boosted ignition, known as energy transfer ignition was introduced by Lucas and Wico-Pacy; the system provided a magneto with an A.C. inductor alternator supplying a special high-tension coil, a split-second timer transferring the A.C. generator output to the high-tension coil at the precise moment of the maximum high-energy A.C. output pulse. This resulted in a boosted high-tension spark caused by the additional voltage rise in the coil windings. This more intense spark was useful for racing and competition work.

Generators and Alternators

The direct-current generators of the Lucas E3H, Lucas C35S, Lucas MC45 and the Miller DVR designs were typical of this type of wound field coil generator which had been used in production for many years in conjunction with a cut-out unit and a voltage regulator unit to control battery charging. The BTH PEC-type generator differed electrically from the normal dynamo in that the field magnets were of the permanent type, which avoided the fluctuating field density, as occurs in the electro-magnetic type of field, and therefore no regulator was required.

An elaboration of the direct-current generator was the Siba Dynastart design which could function either as a generator or a motor, thus providing both battery-charging and engine-starting functions. This design was in the compact and convenient form of a flywheel generator unit with 12 stator coils carried on a fixed backplate, a flywheel-type rotor with built-in armature windings that connect with commutator bars arranged radially on the inner face of the rotor, and a switching system to bring into operation either the generator or the motor function of the unit. A reverse switch could also be used with two-stroke engines to serve as a reverse gear function. Other examples of the Dynastart system were made in Germany by Noris, in Italy by Marelli and in Japan by Mitsubishi. The Bosch pendulum-type starter oscillated instead of rotated the crankshaft in order to start the engine.

The car-type Bendix drive form of engine starter, which turned the engine by a throw-out pinion, was used in the large Harley-Davidson model.

The inductor alternator has the advantage of producing a more constant supply and of being more compact than the direct-current generator, and for these reasons began to be more generally adopted during this period. The alternator consisted of a series of stator coils and pole cores held in a stator frame, with a permanent magnet rotor revolving concentrically within the coils and pole pieces. Designs of alternator were produced by Wipac and Lucas in the RM design; the six stator coils were coupled together in series and connected in parallel for output, one side going to a switch with three lighting positions which brought in one, two or three pairs of coils according to load. Since the output was alternating, it was fed through a silicon rectifier for battery charging.

In 1963, Lucas introduced a new electronic form of voltage control by which a 12-volt battery could be charged through a rectifier from the Lucas RM alternator. This control unit was the Zener diode semi-conductor unit that became conductive in the reverse direction at a critical but pre-determined voltage.

Batteries were generally of the 6-volt 12 ampere-hour type, such as the Lucas PU7E, but with the heavier electrical demands occasioned by electric engine starters, the 12-volt battery began to be more commonly used. The Varley dry battery used very porous positive and negative electrodes and separators, which made it possible for the electrolyte to be absorbed and held in suspension.

A full complex of wiring and switching facilities on the modern motor-cycle provided for full, dip and parking lighting; ignition; stop-light; horn; speedometer light; ignition warning light; fuel-level warning light; and wink-ing direction indicators.

TRANSMISSIONS

The means adopted for transmitting the drive from the engine shaft through a clutch and gearbox to the rear wheel of a motorcycle remained for a consider-able proportion of production models a primary chain and a secondary chain. The shorter primary chain had been enclosed for many years in a case which protected it from dirt and permitted it to operate in a properly lubricated state. With the increasing adoption of the combined engine and gear unit, in which the gearbox case was formed integrally with the engine crankcase, the primary drive consisted in some cases of direct gear drive between a pinion mounted on the crankshaft extension and the driven gear mounted on the gear shaft. In some of the simple ultra-light engine and gear units, such as the

Ariel Pixie (Fig. 14) and the B.S.A. Beagle (Fig. 16), the crankshaft pinion meshed with teeth formed integrally on the peripheries of the free clutch plates. Variations of this standard system of transmission included the adoption of a flexible vee belt for the primary drive running on spring loaded expanding and contracting pulley flanges by means of which the gear ratio was automatically adjusted to load and speed conditions, as used in the Manhurin D.K.W. Hobby and the Concorde designs, and the Mobylette and Raleigh Ultramatic designs; and secondary shaft drive to the rear wheel from the gearbox as used with the Velocette LE200 and Valiant models, the B.M.W. models and the early Lambretta motor scooter. A practical improvement which was increasingly adopted during this period was the readoption of the totally-enclosed case for the secondary chain in the manner of the oil-bath chain case introduced and used extensively before 1914 on Marston Sunbeam bicycles, motorcycles and motorcars.

Clutches were generally manually operated through a hand lever and Bowden cable and were of the spring-loaded single or multi-disc types. The clutch discs were either of the dry type with Ferodo type fabric friction inserts, as used with many English models as well as the American Harley-Davidson models, or the multi-plate wet type operating in oil, as used with Lambretta, Honda and Suzuki models. A simple yet significant modification of clutch design was the centrifugally-loaded clutch which automatically engaged as the engine revolutions increased. These simple automatic clutches were generally used with small motorcycles such as the Lambrettino, Suzuki 50, Mobylette and Raleigh cyclemotors and the Triumph T-10 motor scooter.

Simple two-speed gears were used with some of the simplest cyclemotor models, such as the early N.S.U. Quickly design. Three and four-speed gearboxes with constant-mesh gears were still used as separate units, although the growing tendency to incorporate the gear train in an extension of the crankcase to form a compact engine and gear unit has already been noted. Both the separate and the unit gearboxes incorporated positive foot-operated returnshift gear-changing mechanism and a kick-starter device. More rarely, electric engine starting was effected by means of a Siba Dynastart-type unit or, as in the case of the Harley-Davidson Electra Glide model, a Bendix-type motor and throw-out pinion.

Some of the more advanced motorcycle designs of small cylinder capacities, high specific outputs and high crankshaft speeds which, by virtue of their essential design, had narrow power bands (i.e. the useful power being developed only over a small band of the higher crankshaft speeds), required the flexibility provided by a greater number of available speed-gear ratios, so that five, six and even seven-speed gearboxes were provided for racing and, in some cases, also for production motorcycles. The miniature racing machines of from 50 c.c. to 125 c.c. cylinder capacities, in particular, which had extremely narrow power bands at very high crankshaft speeds, required as many as 18 speed-gear ratios to maintain these miniature engines within their useful power bands over a widely varying road course such as the Isle of Man Tourist Trophy course. This high number of gear ratios was achieved by providing a secondary gearbox as an overdrive unit which, in conjunction with a normal four-speed gearbox, multiplied the standard number of gear ratios by the number of gear ratios of the secondary box. A four-speed gearbox, for instance, associated with a three-speed secondary gearbox, would provide twelve separate close-ratio gears.

Some simple automatic gear changing systems were used on certain ultra-light motorcycle designs. The Uher system used with Manuhurin, D.K.W. Hobby and Concorde designs employed spring-loaded split driving and driven pulleys connected by a flexible vee belt; when the load on the driving pulley increased, the pulley flanges were pushed further apart and the belt sank to a smaller effective pulley diameter, thus lowering the gear ratio. At the same time, the load on the driven pulley diminished because of the slackening of the belt, and the spring-loaded flanges tended to come together, thus increasing the effective diameter of the driven pulley. These complementary actions resulted in a lowering of the overall gear ratio. A lessening of the road load, on the other hand, had the converse effect and the overall gear ratio was raised accordingly. The two pulleys were thus constantly adjusting themselves to varying load conditions to provide optimum operational gear ratio settings. The Mobylette-Raleigh system used only one spring-loaded expanding and contracting pulley on the engine shaft, and the engine was spring-mounted on a pivot which permitted it to be moved either nearer or further from the countershaft pulley to maintain effective belt tension as the gear ratio varied.

FRAMES AND SUSPENSIONS

The long established diamond type of frame, evolved for the safety bicycles in the 1890's, continued to be used in hardly modified forms for many of the medium and heavier designs of motorcycle. The more straightforward versions still used the single tube original type with the front down tube, the engine crankcase being incorporated by means of lugs and brackets in the lower apex of the diamond frame. A variation of this original form intended to provide greater strength and stiffness without significant increase of weight was the duplex or cradle form of frame, which was used not only for the heavier motorcycles intended for use with a sidecar, but was also used with some of the lighter solo models. An interesting variation of the original diamond frame, introduced as long ago as 1901 for use with early Humber and P. & M. motorcycles, and continued in production to this present period in the form of the Panther design, incorporated the inclined single-cylinder engine to serve as the front down tube of the diamond frame.

More fundamental changes in motorcycle frame design were adopted during this period in the lightweight categories of motorcycle, particularly the new forms of cyclemotor, motor scooter and ultra-light motorcycle proper, as had been foreshadowed in the tentative designs of these categories which had appeared during the previous two decades. The most general form for the two former categories was the single spine or backbone frame, which was generally formed from welded steel pressings, incorporating lugs and brackets for the attachment of steering head and rear forks, the latter being articulated to serve as sprung rear suspension. This welded spine frame was cheap and simple to make in mass production and substantially simplified the make-up of the small motorcycle. Cyclemotors generally used this single spine construction in underhung form, such as adopted in the Honda 50 (Plate 15), the

Suzuki 50 (Plate 15), the Raleigh Ultramatic, the N.S.U. Quickly and many others of this category. The ultra-light motorcycles proper, on the other hand, generally adopted the overhung form of spine frame as was used in such designs as the Suzuki Sports 50, Suzuki 120, Kreidler Florette (Plate 17) and many other models of this category. For the Raleigh Wisp (Plate 15) cyclemotor there was adopted a tubular underhung single spine form of frame which stemmed from the Spencer Moulton and Raleigh types of current bicycle frames, which in turn had been foreshadowed by the cross-frame safety bicycle frame of 1886. The Little Honda P50 and Clark Scamp miniature cyclemotors with their 49 c.c. capacity motorized rear wheels, used a spine frame of the same general simplicity constructed respectively of either of welded steel pressings or steel tubes. The spine frames of the Raleigh Wisp, Clark Scamp and the Little Honda cyclemotors represent the simplest form of frame possible to use for motorcycle construction.

The open frame generally adopted for motor scooters was either of the single tubular spine kind or the duplex tubular cradle kind with lug attachments to form the platform shield, and of specialized shape to accommodate steering head, engine attachment and articulated rear sprung forks.

Sprung suspension at both front and rear ends of almost all forms of motor-cycles was now general practice. In the simplest forms, such as the Raleigh Wisp, oversize tyres were the sole means of absorbing road shocks; the Little Honda had in addition a simple form of cantilever sprung front wheel. This same form of front wheel springing, with either leading or trailing cantilever arms, was used for many cyclemotor, motor scooters and ultra-light motor-cycle designs, but these categories also generally incorporated articulated rear wheel forks supported by spring-suspension units. Telescopic sprung front forks were more suitable for the larger forms of motorcycle, even up to the largest sizes such as the 1200 c.c. Harley-Davidson, whose telescopic front forks incorporated both hydraulic suspension cylinder and rebound coil springs to meet the heavier load conditions entailed. A coil-sprung saddle pillar was also used to increase riding comfort. Rear suspensions in general consisted of articulated rear wheel forks supported by a pair of sprung shock absorbing units attached between the rear ends of the fork extensions and a prolongation of the upper main frame. In some cases, such as with the Velocette LE200 design, rear spring tension could conveniently be adjusted to match different load and road conditions. A variety of specialized frame designs for racing motorcycles was evolved, intended to save weight and provide better riding and road stability characteristics.

BRAKES

The ever-increasing emphasis on road safety required a parallel need for efficient braking mechanism, not only to deal with increased road performance but also to ensure as far as possible safer motorcycling for an increasing number and variety of motorcyclists. All forms of motorcycles now had efficient manually-operated front and foot-operated rear wheel braking mechanisms. Brakes were of the internal-expanding drum type, of substantial dimensions to ensure adequate braking effort without overheating and tendency to fade. Brake drums were sometimes made of light alloy, finned to disperse heat, and fitted with cast iron liners against which the Ferodo-lined internal-expanding brake shoes could bear. Additional aids to braking efficiency were the hydraulic rear wheel brake fitted to the 1200 c.c. Harley-Davidson, and the front wheel disc brake used on the Lambretta SX200 motor scooter; these were indications of future progress towards more effective braking provision for general production motorcycles.

TYRES

A wide variety of tyres by different first-class manufacturers was available for use with motorcycles of all kinds and for all purposes, from the small 3·50 x 8 in. size for motor scooters to the 4·5 x 19 in. size for the heaviest motorcycles. High-hystersis rubber compounds that gave greatly improved road grip were now used in commercial production, together with a large variety of tread patterns providing the greatest efficiency for different road surface conditions, such as fairly smooth patterns for normal surfaces, more scientifically complex patterns having a self-draining squeegee action for wet surfaces to inhibit tendency to aquaplaning, and wide and deep patterns for trials and scrambles to give maximum grip on muddy surfaces. Ultimate cornering adhesion and very high racing speeds require a shallow tyre tread, wrapping well round the tyre wall. On the other hand, sidecar machines which do not lay over on cornering require a broad tread that retains full contact with the road surface. Considerable advance in both natural and synthetic rubber compounds, cotton, rayon and nylon-ply construction, and the scientifically designed tread patterns already mentioned during this period resulted in a total gain of 40 per cent in road grip. This improvement in stopping distance and cornering stability provided greater operational safety, and enabled tyres of better performance and longer life to be designed and made.

SIDECARS

The former popularity of the motorcycle sidecar combination (Plate 21) as an economical form of family transport declined during the previous two decades with the advent of the efficient and economical saloon light car and a rising standard of living which permitted its purchase by many classes of people. Sidecar design and manufacture, however, did continue on a diminishing scale for both family and commercial transport purposes. Such sidecar combinations of this latest period included the largest size associated with heavy American motorcycles equipped with a reverse gear to facilitate the handling of so heavy a combination; the medium size of sidecar used in conjunction with motorcycles described in Chapter 9, with one or two seats and full weather protection; open sports sidecars with windscreens; and the smallest sizes which were specially developed for use with the latest generation of motor scooters which, as solo mounts, were capable of a high performance with two persons, and so were capable of hauling a small light sidecar usually for commercial transport purposes. Sidecar bodies were constructed of resin-bonded plywood, aluminium-alloy panels or, in the more expensive designs, moulded glass fibre, the latter being capable of being formed into graceful, streamlined shapes. Bodies were mounted on a light tubular frame having a three-point attachment to the motorcycle frame, and the single wheel was spring on rubber in torsion. Some of the old established sidecar manufacturers still in production at this period included such names as Blacknell, Canterbury, Garrard and Watsonian.

Bibliography

A small selection of the more important books (most of which are kept in the Science Museum Library) dealing with the history, design and construction of motorcycles.

BEAUMONT, L. WORBY	*Motor Vehicles and Motors* (vol. 1.)	1900
	(vol. 2.)	1906
CAUNTER, C. F.	*The Two-Cycle Engine*	1932
DAVISON, G. S.	*The Story of the T.T.*	1952
DUNCAN, H. O.	*The World on Wheels* (2 vols.)	1928
CORBISHLEY, H.	*Motorcycles and Sidecars*	1927
FARR, F. L.	*Mo-peds and Scooters*	1960
HIGGINS, L. R.	*Britain's Racing Motor Cycles: 1897–1952*	1952
HINGSTON, I. R.	*Scooters and Mopeds*	1958
JACKMAN, W. J.	*A.B.C. of the Motor Cycle* (U.S.A.)	1910–1916
JONES, B. E.	*Motorcycles*	1915
'IXION'	*Motor Cycle Reminiscences*	1920
	Further Motor Cycle Reminiscences	1926
	Motor Cycle Cavalcade	1950
ILIFFE LTD.	*The Motor Cycle*	from 1903
	Two-Stroke Motor Cycles (various editions)	from 1921
IRVING, P. E.	*Motorcycle Engineering*	1964
LOW, A. M., Dr.	*Two-Stroke Engines*	1915
PAGÉ, VICTOR W.	*Motorcycles, Sidecars and Cyclecars* (3rd edition) (U.S.A.)	1914
'PHOENIX'	*The Motor Cyclists' Handbook* (4th edition)	1914
SAUNIER, BAUDRY DE DOLLFUS, CHARLES	*Histoire de la Locomotion Terrestre* (vol. 2)	1936
SHELDON, J.	*Veteran and Vintage Motorcycles*	1961
STEVENS, J.	*Scootering*	1962
TEMPLE PRESS LTD.	*Motor Cycling*	1902–1903 and from 1909
	Motor Cycling Manual (Several editions)	from 1912
WALFORD ERIC, W.	*Early Days in the Motor Cycle Industry*	1931
WILSON, A. J.	*Motor Cycles and How to Manage Them:* (Several editions)	from 1903
ZÉROLO, M.	*Motorcyclettes et Tricars*	1916
TRAGATSCH, E.	*The World's Motorcycles: 1894–1963*	1964

Index

A.B.C. Motorcycle, 33, 41, 49 54,
A.B.C. Skootamota, 52, 78, 81,
Ace motorcycle, 50, 61,
Adler motor scooter, 84, 85,
Aermacchi motorcycle, 89,
Aero Caproni Trento motorcycle, 70,
A.J.S. motorcycle, 32, 42, 49, 59, 67, 68, 69, 89, 92, 95, 96,
Alcyone motorcycle, 18,
Allard motor tricycle, 9,
Alldays & Onions motorcycle, 18,
A.M.A.C. carburettor, 13, 25, 37, 53,
Amal carburettor, 69, 101,
Anglian gear, 26,
Antoine motorcycle, 18,
Anzani engine, 18, 20, 49,
Anzani motorcycle, 18,
Ariel motorcycle, 9, 23, 24, 31, 61, 67, 68, 87, 89, 96, 104,
Armstrong-Triplex hub gear, 40,
A.S.L. motorcycle, 41,
Aspin engine, 62,
Aster motor tricycle, 12,
Auto-Cycle Club, 29,
Auto-Cycle Union, 29,
'Autolube' lubrication, 101,
Autoglider motor scooter, 81,
Autoped scooter, 52,
Autowheel, 36, 52, 78,

Barr & Stroud engine, 48,
Barry rotary engine, 21,
Barter, J. F., 17, 21,
Bat motorcycle, 27, 32, 41, 42,
Bayliss-Thomas motorcycle, 18,
Beardmore-Precision motorcycle, 51,
Beeston motor tricycle, 7,
Bendix starter, 97, 153,
Benelli motorcycle, 90, 94,
Benz engine, 12,
Bercley engine, 21,
Berini cyclemotor, 74,
Bianchi motorcycle, 84, 88, 94,
Bichrone engine, 23,
Binetta cyclemotor, 80,

Binks carburettor, 38, 53,
Binks engine, 21,
Bins carburettor, 101,
Binz motor scooter, 83,
Black Prince engine, 51,
B.M.W. motorcycle, 40, 60, 64, 75, 89, 95,
Bock & Hollander motorcycle, 18,
Bonniksen speedometer, 43,
Bosch high-tension magneto, 18, 25, 37, 53,
Bosch starter generator, 85,
Bosch pendulum starter, 103,
Bosch, Robert, 12, 36, 101,
Boudeville, André, 12,
Bouton, Georges, 2, 8,
Bowden cable, 25, 104,
Bowden gear, 41,
Bradbury motorcycle, 18, 19 23, 51, 32, 53,
Bradshaw engine, 33, 48, 49,
Brampton spring fork, 42, 64,
Brée engine, 23,
British Motor Traction Company, 16,
Brooks saddle, 28, 42,
Brough motorcycle, 33, 49, 59, 62,
Brown & Barlow carburettor, 13, 25, 37, 53,
Brown motorcycle, 23,
B.S.A. hub gear, 40,
B.S.A. motorcycle, 27, 31, 37, 46, 58, 59, 67, 68, 88, 89, 92, 95, 96, 104,
B.S.A. motor scooter, 84, 86,
B.S.A. 'Winged-Wheel' motor-wheel, 74,
B.T.H. 'Mag-generator', 56,
B.T.H. magneto, 37, 53, 101,
B.T.H. 'Sparklite', 55,
Buchet engine, 18, 20,
Bultaco motorcycle, 87, 89,
Burney and Blackburne engine, 31, 48, 53,
Burt-McCollum engine, 48,
Butler, Edward, 4, 22,

Butler 'Inspirator' spray carburettor, 4, 13,
Butler 'Petrol-Cycle', 4, 12, 22,

Capri motor scooter, 84,
Capri cyclemotor, 80,
Capriolo motorcycle, 87, 88,
C.A.V. 'Dynamag', 55,
C.A.V. magneto, 25, 37, 53,
Chater-Lea gear, 26, 32,
Chater-Lea motorcycle, 42, 64,
Clark Scamp cyclemotor 79, 106,
Claudel-Hobson carburettor, 54,
Clément motorcycle, 16, 18, 20, 27,
Clerk two-stroke engine, 4, 22, 23,
Cleveland motorcycle, 50,
Clyno motorcycle, 32, 35, 38, 40, 41, 42, 44, 46,
Concorde motor scooter, 82, 104, 105,
Connaught motorcycle, 35,
Cooper-Stewart speedometer, 43,
Copeland, L. D., 2,
Corah engine, 32, 62,
Corbin speedometer, 43,
Corgi motor scooter, 72,
Coventry-Eagle motorcycle, 64,
Coventry Motor Company's ladies' motorcycle, 11,
Coventry-Victor engine, 49,
Cowey speedometer, 43,
Cox-Atmos carburettor, 54,
Cox carburettor, 37, 54,
Cross engine, 62,
'Cucciolo' cyclemotor, 74,
Curtiss motorcycle, 19,
'Cyclemaster' motor-wheel, 74,
CZ motorcycle, 88, 93, 95,

Daimler engine, 3, 12, 13, 18,
Daimler, Gottlieb, 3,
Daimler motorcycle, 4,
Daimler surface vaporiser, 3,
Dalifol steam motorcycle, 2,
Dalm engine, 35, 51,
Day two-stroke engine, 22,
Dayton engine, 36, 78,
De Dion-Bouton carburettor, 13, 24,

De Dion-Bouton engine, 7, 8, 13, 15, 18, 25,
De Dion-Bouton ignition system, 12, 15, 25,
De Dion-Bouton motor tricycle, 7, 8, 9, 11,
De Dion, Count Albert, 2, 8,
Degory carburettor, 54,
De Lissa valve, 31,
Dell'Orto carburettor, 101,
Desmodromic valve gear, 88, 90,
Diamond engine, 41,
Dixie magneto, 37,
DKR motor scooter, 86,
D.K.W. motorcycle, 52, 62, 63, 71, 72, 89, 105,
D.K.W. motor scooter, 80, 82, 83,
D.M.W. motorcycle, 89, 92,
Douglas gear, 40,
Douglas motorcycle, 21, 32, 33, 40, 46, 49, 60,
Druid spring fork, 27, 41, 64,
Ducati motorcycle, 70, 83, 89, 90, 99,
Dufaux, H. and A., 16,
Dunelt motorcycle, 51,
Dunlop, J. B., 14, 42,
Dürkopp motorcycle, 18, 21, 80,
Dürkopp motor scooter, 85,

Eadie motor tricycle, 9,
Earles spring fork, 75,
Edmund motorcycle, 41,
Ehrlich engine, 72,
E.I.C. magneto, 53,
Eisemann magneto, 25, 37,
Elgin engine, 23,
Ellehamer motor scooter, 78,
E.M.C. motorcycle, 89,
Enfield gear, 26,
Enfield motor tricycle, 9,
Eta motorcycle, 50,
Excelsior motorcycle, 31,
Excelsior engine, 71, 92, 95,
Excelsior (U.S.), 32,

Fafnir engine, 18, 31,
Fafnir gear, 25,
Fallolite lamp, 55,

Fée (Fairy) motorcycle, 17, 21,
Fielbach motorcycle, 39,
F.N. carburettor, 24,
F.N. motorcycle, 18, 21, 27, 28, 32, 33, 42, 44, 49, 50, 54, 64,
F.N. cyclemotor, 80,
Foster-Ital motorcycle, 87,
Francis Barnett motorcycle, 64,
Franklin, C.B., 47,
F.R.S. dynamo, 43,
'Fuelmix' lubrication, 101,

Garelli cyclemotor, 80, 84,
Garelli motorcycle, 87,
Garrard, C. R., 16,
Garrard forecar, 10,
Garrard sidecar, 108,
Gibson motorcycle, 11,
Gilera motorcycle, 62, 64, 70, 88, 94, 96,
Gilera cyclemotor, 80,
Gladiator motor tricycle, 11,
Godfrey, O. C., 31,
Grado gear, 39,
Graham, W. G., 10,
Green engine, 22,
Greeves motorcycle, 89, 90, 92, 93,
Gregoire motorcycle, 18,
Griffon motorcycle, 18, 20,
Guzzi motorcycle, 59, 61, 70, 72, 80, 87,

Harley-Davidson hub gear, 40,
Harley-Davidson motorcycle, 19, 20, 25, 32, 42, 47, 49, 96, 97, 103, 104, 106, 107,
Harley-Davidson motor scooter, 85,
Harper motor scooter, 52, 81,
Hedstrom carburettor, 25,
Hedstrom, Oscar, 19,
Heinkel motor scooter, 85,
Hendee, George M., 19,
Hendee Manufacturing Company, 19, 44, 47,
Henderson motorcycle, 34, 50,
Hercules cyclemotor, 80,
Hildebrand and Wolfmuller motor-cycle, 5, 12, 13,

Holden motorcycle 6, 12, 13,
Honda cyclemotor, 80, 81,
Honda engine, 80, 87, 88, 90, 99,
Honda motorcycle, 87, 88, 89, 90, 92, 94, 95, 96, 99, 104, 105, 106,
Honold, G., 12,
Hooydonk, J. Van, 10,
Humber ladies' motorcycle, 11,
Humber motorcycle, 11, 17, 19, 24, 32, 49, 105,
Humber 'Olympia Tandem' tricar, 10,

Ilo motorcycle, 71,
Imperia motorcycle, 64,
Indian electric system, 43,
Indian gear, 40,
Indian motorcycle, 19, 20, 25, 27, 31, 32, 38, 39, 40, 41, 42, 43, 46, 47, 49, 50, 61,
Injectolube lubrication, 101,
Innocenti Company, 81,
Iso motorcycle, 72,
Ixion engine, 23,

James motorcycle, 18, 31, 38, 41 89,
James cyclemotor, 80,
J.A.P. engine, 18, 19, 20, 31, 32, 48, 49, 59,
Jawa motorcycle, 71, 83, 87, 89, 90, 93, 94, 95,
Jawa cyclemotor, 80,
Jawa engine, 71,
Jebb scooter, 78,
J.E.S. engine, 36, 78,
Jones speedometer, 43,
Juckes gear, 41,

Kawasaki motorcycle, 88, 89, 93,
Kei-Hin carburettor, 101,
Kelecom, Paul, 18, 50,
Kenilworth motor scooter, 52,
Kerry motorcycle, 23,
Kingsbury motor scooter, 81,
Krebs carburettor, 13, 24,
Kreidler cyclemotor, 80, 85, 106,
Kreidler motorcycle, 87,

Lagonda forecar, 10,
Lambretta motor scooter, 72, 73, 82, 83, 84, 85, 86, 104, 107,
Lambrettino cyclemotor, 80, 104,
Lamps, oil, acetylene, electric, 14, 42,
Laurin & Klement motorcycle, 18, 21,
Lawson, H. J., 7,
Lea-Francis motorcycle, 32, 41,
Lenoir, E., 12,
Levis motorcycle, 35,
Lloyd engine, 33,
L.M.C. Auto-Varia gear, 39,
Locomotives on Highways Act of 1896, 7,
Lodge sparking plugs, 25,
Lohmann cyclemotor, 74,
Longuemare carburettor, 13, 24, 37,
Longuemare-Hardy carburettor, 54,
'Lubematic' lubrication, 101,
Lucas generator, 102, 103,
Lucas ignition, 101, 102,
Lucas 'Magdyno', 55,
Lucas lamps, 14, 56,
Lycett saddle, 42,

Mabon gear, 39,
MacEvoy motorcycle, 59,
M.A.G. engine, 31, 32,
'Maglite' lighting system, 43,
Maico motor scooter, 85, 86,
Mangin reflector, 42,
Manurhin motor scooter, 83, 104,
Manurhin automatic transmission, 105,
Marelli flywheel starter, 84, 103,
Marsh motorcycle, 19,
Matchless motorcycle, 18, 23, 32, 44, 58, 60, 61, 68, 89, 92, 95, 96,
Maybach spray carburettor, 13,
Maybach, Wilhelm, 4,
Meek steam motorcycle, 2,
Megola motorcycle, 50,
Merkel motorcycle, 19,
Michaux-Perreaux steam motorcycle, 1,
Militaire motorcycle, 50,
Miller generator, 102,
Millet motorcycle, 5,

Mills & Fulford sidecar, 10,
Minerva engine, 16, 18, 20, 23, 27, 31, 52,
Mira dynamo, 43,
Mitsubishi flywheel starter, 103,
M.L. 'Maglita', 56,
M.M.C. motorcycle, 15, 18,
M.M.C. motor tricycle, 9,
Mobylette automatic transmission, 104,
Mobylette cyclemotor, 79, 83, 104,
Mondial motorcycle, 88,
Montesa motorcycle, 87, 88,
Montgomery engine, 33,
'Mosquito' cyclemotor, 74,
Motorbi cyclemotor, 83, 84,
Moto-Guzzi motorcycle, 89, 96,
Moto-Guzzi cyclemotor, 80,
Moto-Guzzi engine, 96,
Motor Cycle, The, 29,
Motor Cycling, 28,
Moto Reve motorcycle, 32,
Motor Museum, 2,
Motosacoche motorcycle, 32,
M.V. motorcycle, 70, 90, 94 96,
MZ motorcycle, 87, 88, 89, 90, 100,

Ner-a-car motorcycle, 52,
New Era motorcycle, 19,
New Hudson motorcycle, 18, 80,
New Imperial motorcycle, 59,
Newton, A. V., 12,
Nimbus motorcycle, 61,
Norman cyclemotor, 80,
Noris flywheel starter, 103,
Norton motorcycle, 31, 46, 48, 58, 59, 67, 68, 89, 92, 95, 96,
N.S.U. gear, 23, 36, 39,
N.S.U. motorcycle, 18, 20, 32, 41, 42, 70, 71, 79, 89, 96,
N.S.U. cyclemotor, 79, 87, 104, 106,
N.S.U. motor scooter, 82, 83, 85,
N.U.T. motorcycle, 32,

O.E.C. motorcycle, 64,
Opel, F., 50,
Oscar motor scooter, 73,

Ossa motorcycle, 89,
Otto, Dr. N. A., 3,

P. & M. gear, 17, 26, 38, 40,
P. & M. motorcycle, 27, 31, 44, 46, 49, 67, 96, 105,
Parilla motorcycle, 92,
Panther motorcycle, 92,
Parkyns-Bateman steam tricycle, 2,
Pebok engine, 18,
Peco engine, 35,
Pennington motorcycle, 6,
Perks & Birch, 11.
Peugeot motorcycle, 18, 20, 31, 32,
Peugeot cyclemotor, 80,
Peugeot motor scooter, 85,
Phelon & Moore, 17, 24, 27,
Philipson gear, 39,
Phoenix motorcycle, 10,
Pheonix motor scooter, 85,
Phoenix 'Trimco' forecar, 10,
Piatti motor scooter, 85,
Piaggio Company, 81,
Pierce-Arrow motorcycle, 33, 39,
Pope motorcycle, 19, 32,
Posi-Force lubrication, 101,
Powell & Hanmer dynamo, 42,
Power Pak cyclemotor, 74,
Precision engine, 31 32, 41,
Premier motorcycle, 31, 32, 35,
Prestwich, J. A., 18, 19, 31, 32, 48, 70,
Progress motorcycle, 18,
Püch motorcycle, 18, 20, 32, 52, 64, 71, 72, 87, 89,
Püch cyclemotor 80, 84, 85, 87,
Pullin C. G., 31,
Pullin-Groom motorcycle 51,

Quadrant forecar, 10,
Quadrant motorcycle, 18, 19, 23, 35,

Raleigh motorcycle, 15, 18,
Raleigh cyclemotor, 79, 106,
Raleigh motor scooter, 84, 104, 106,
Raleigh automatic transmission, 79, 105,
R.A.C. motorcycle, 93,
Reading-Standard motorcycle, 32,

Redrup C. R., 21, 34, 50,
Renaux motor tricycle, 9,
Rex motorcycle, 18, 19, 23, 32, 35,
Ricardo, H. R., 48,
Rich detachable tube, 42,
Riley forecar, 10,
Robinson & Price motorcycle, 17,
Roc gear, 23, 25, 40,
Roc motorcycle, 23, 35,
Ro-Monocar, 53,
Roots, J. D., 5, 22,
Roper, S. H., 2,
Rover forecar, 10, 19,
Rover motorcycle, 19,
Royal Enfield motorcycle, 15, 18, 19, 32, 35, 38, 40, 51, 58, 59, 67, 89, 92, 95, 96,
Royal motorcycle, 19,
Rudge Multi gear, 39,
Rudge motorcycle, 27, 30, 31, 48,
Rumi motor scooter, 85,
Runbaken magneto, 53,
Runbaken 'Magneto-Lite', 55,
Ruthardt magneto, 25,

Sachs engine, 80,
Sarolea motorcycle, 18, 31,
Saxon spring fork, 42,
Schebler carburettor, 25,
Scheibert engine, 23,
Schickel motorcycle, 36,
Schliah engine, 52,
Schnuerle loop-scavenge 52, 62, 63, 71,
Schulte, M. J., 23,
Scott, A. A., 22, 27, 34,
Scott 'Cyc-Auto', 73,
Scott gear, 40,
Scott motorcycle, 22, 27, 34, 38, 42, 46, 51, 55, 62, 63, 64, 93,
Senspray carburettor, 37,
Serpollet, Léon, 2,
Service belt, 38,
Sharpe spring frame, 27,
Shaw motorcycle, 11,
Siba Dynastart, 92, 95, 102, 104,
Sidecars, 10, 43, 56, 107,

Simms-Bosch low-tension magneto, 12,
Simms engine, 18,
Simms, F. A., 12,
Simms high-tension magneto, 12, 37,
Singer forecar, 10,
Singer motorcycle, 18,
Singer motorwheel, 11,
Smith speedometer, 43,
Solex cyclemotor, 74, 79,
Spacke motorcycle, 32,
Sphinx sparking plugs, 25,
Splitdorf magneto, 37,
Stanger engine, 51,
Stewart-Precision carburettor, 38,
Sturmey-Archer hub gear, 40, 46,
Sun engine, 51,
Sunbeam motorcycle 30, 31, 32, 58, 41, 42, 46, 48, 55, 68, 75, 104,
Suzuki cyclemotor, 80, 88,
Suzuki motorcycle, 87, 89, 90, 94, 100, 104, 106,
Swift motor tricycle, 9,

T.A.C. Wilkinson motorcycle, 33, 34,
Tansad pillion, 57,
T.D.C. engine, 51,
Terrot motor scooter, 85,
Thomas motorcycle, 19,
Thompson-Bennett magneto, 37, 53,
Tourist Trophy Race, 24, 30, 32, 34, 35, 39, 44, 48, 57, 58, 59, 60, 61, 64, 75, 78, 94, 104,
Triumph hub clutch, 40,
Triumph motorcycle, 17, 18, 23, 24, 28, 30, 31, 33, 35, 40, 42, 46, 51, 60, 67, 70, 89, 92, 95, 96,
Triumph motor scooter, 84, 86, 104,
Triumph T.W.N. motorcycle, 72,
TWN motor scooter, 85,
Trojan 'Minimotor', 74,
Truffault spring fork, 27, 75,
Tyres, pneumatic, 14, 42, 57, 64, 107,

U.H. Magneto, 25, 37,
Uher transmission, 82, 105,
Union engine, 51,

Varley battery, 103,
Vaurs carburettor, 13, 24,
Veeder distance counter, 43,
Veloce engine, 32, 41,
Velocette motorcycle, 35, 48, 51, 54, 59, 67, 68, 75, 89, 92, 106,
Velocette motor scooter, 83, 86,
'Vélocipédraisiavaporianna', 1,
Verdon-Roe, Sir Alliott, 53,
Vespa motor scooter, 72, 81, 83, 85,
Victoria motorcycle, 62, 64,
Victoria motor scooter, 83,
Villiers engine, 32, 35, 41, 51, 53, 62, 63, 71, 72, 78, 80, 86, 90, 92, 101,
Villiers flywheel magneto, 51, 102,
Villiers hub clutch, 40,
Villiers-Mills carburettor, 54, 101,
Vincent-H.R.D. motorcycle, 59,
Vincent motorcycle, 66, 68, 69,
Vindec gear, 26,
Vostok engine, 95,

Wagner motorcycle, 19,
Wall Autowheel, 36, 52, 78,
Wanderer motorcycle, 32, 41,
Watawata belt, 27,
Watford speedometer, 43,
Watsonian sidecar, 108,
Webb spring fork, 55, 64,
Werner, Michel and Eugéne, 15, 18, 20,
Werner motorcycle, 13, 15, 16, 17, 18, 21
White and Poppe engine, 18, 19,
Whittle belt, 38,
Wico-Pacy ignition coil, 102, 103,
Wilkinson motorcycle, 33, 39, 41,
Williamson motorcycle, 33,
Wooler motorcycle, 35, 41, 49, 69,

Xl-all saddle, 42,

Yale motorcycle, 19,
Yamaha motor scooter, 80,
Yamaha motorcycle, 80, 87, 88 89, 90, 93, 94, 100,

Zenette motorcycle, 27,
Zener diode, 103,
Zenith 'Gradua' gear, 39,
Zenith motorcycle, 32,

Zündapp motorcycle, 52, 62, 64, 70,
Zündapp motor scooter, 73, 85, 86,
Zündapp cyclemotor, 80,